Freeman, Edward

Atlas to the Historical Geography of Europe from Homeric Greece until 1900

Freeman, Edward

Atlas to the Historical Geography of Europe from Homeric Greece until 1900

ISBN: 978-3-86741-378-7

Auflage: 1
Erscheinungsjahr: 2010
Erscheinungsort: Bremen, Deutschland

© Europäischer Hochschulverlag GmbH & Co KG, Fahrenheitstr. 1, 28359 Bremen (www.eh-verlag.de). Alle Rechte beim Verlag und bei den jeweiligen Lizenzgebern.

Bei diesem Titel handelt es sich um den Nachdruck eines historischen, lange vergriffenen Buches aus dem Verlag Longmans, Green & Co, London/New York/Bombay (1903). Da elektronische Druckvorlagen für diese Titel nicht existieren, musste auf alte Vorlagen zurückgegriffen werden. Hieraus zwangsläufig resultierende Qualitätsverluste bitten wir zu entschuldigen.

Freeman, Edward
**Atlas to the Historical Geography of Europe
from Homeric Greece until 1900**

ATLAS

TO THE

HISTORICAL GEOGRAPHY

OF

EUROPE

BY

EDWARD A. FREEMAN, D.C.L., LL.D.

FORMERLY REGIUS PROFESSOR OF MODERN HISTORY IN THE UNIVERSITY OF OXFORD

THIRD EDITION

EDITED BY

J. B. BURY, M.A., D.Litt., LL.D.

REGIUS PROFESSOR OF MODERN HISTORY IN THE UNIVERSITY OF CAMBRIDGE

LONGMANS, GREEN, AND CO.
39 PATERNOSTER ROW, LONDON
NEW YORK AND BOMBAY
1903

All rights reserved

LIST OF MAPS.

Nos.

I. HOMERIC GREECE AND THE NEIGHBOURING LANDS.

II. GREECE AND THE GREEK COLONIES.

III. GREECE IN THE FIFTH CENTURY B.C.

IV. THE LANDS ROUND THE ÆGÆAN AT THE BEGINNING OF THE KLEOMENIC WAR, c. B.C. 227.

V. DOMINIONS AND DEPENDENCIES OF ALEXANDER, c. B.C. 323.

VI. KINGDOMS OF THE SUCCESSORS OF ALEXANDER, c. B.C. 300.

VII. ITALY BEFORE THE GROWTH OF THE ROMAN POWER.

VIII. THE MEDITERRANEAN LANDS AT THE BEGINNING OF THE SECOND PUNIC WAR.

IX. THE ROMAN DOMINIONS AT THE END OF THE MITHRIDATIC WAR, B.C. 64.

X. THE ROMAN EMPIRE AT THE DEATH OF AUGUSTUS, A.D. 13.

XI. THE ROMAN EMPIRE UNDER TRAJAN, A.D. 117.

XII. THE ROMAN EMPIRE DIVIDED INTO PREFECTURES.

LIST OF MAPS.

Nos.

XIII. EUROPE IN THE REIGN OF THEODORIC, c. A.D. 500.

XIV. EUROPE AT THE DEATH OF JUSTINIAN, A.D. 565.

XV. EUROPE AT THE END OF SEVENTH CENTURY, A.D. 695.

XVI. GREATEST EXTENT OF THE SARACEN DOMINIONS.

XVII. EUROPE IN THE TIME OF CHARLES THE GREAT, A.D. 814.

XVIII. THE WESTERN EMPIRE AS DIVIDED AT VERDUN, A.D. 843.

XIX. THE WESTERN EMPIRE AS DIVIDED A.D. 870.

XX. DITTO DITTO A.D. 887.

XXI. CENTRAL EUROPE, c. A.D. 980.

XXII. CENTRAL EUROPE, A.D. 1180.

XXIII. DITTO A.D. 1360.

XXIV. DITTO A.D. 1460.

XXV DITTO A.D. 1555.

XXVI. DITTO A.D. 1660.

XXVII. DITTO A.D. 1780.

XXVIII. DITTO A.D. 1801.

XXIX. DITTO A.D. 1810.

XXX. DITTO A.D. 1815.

XXXI. DITTO A.D. 1860.

XXXII. DITTO A.D. 1871.

LIST OF MAPS.   vii

Nos.
XXXIII. BOUNDARIES OF FRANCE, A.D. 1555, 1715, 1791, 1871.

XXXIV.   SOUTH-EASTERN EUROPE, c. A.D. 910.

XXXV.   DITTO   DITTO   c. A.D. 1000.

XXXVI.   DITTO   DITTO   c. A.D. 1040-1070.

XXXVII.   DITTO   DITTO   c. A.D. 1105.

XXXVIII.   DITTO   DITTO   c. A.D. 1180.

XXXIX.   DITTO   DITTO   c. A.D. 1210.

XL.   DITTO   DITTO   c. A.D. 1340.

XLI.   DITTO   DITTO   c. A.D. 1354-1358.

XLII.   DITTO   DITTO   c. A.D. 1401.

XLIII.   DITTO   DITTO   c. A.D. 1444.

XLIV.   DITTO   DITTO   c. A.D. 1464.

XLV.   DITTO   DITTO   c. A.D. 1672.

XLVI.   DITTO   DITTO   c. A.D. 1700.

XLVII.   DITTO   DITTO   c. A.D. 1727.

XLVIII. SOUTH-EASTERN EUROPE, A.D. 1861.

XLIX.   SOUTH-EASTERN EUROPE, A.D. 1881.

L.   THE BALTIC LANDS, c. A.D. 1000.

LI.   DITTO   c. A.D. 1220.

LII.   DITTO   c. A.D. 1270.

LIII.   DITTO   c. A.D. 1350-60.

LIST OF MAPS.

Nos.
LIV. THE BALTIC LANDS, c. A.D. 1400.

LV. DITTO c. A.D. 1478.

LVI. DITTO c. A.D. 1563.

LVII. DITTO c. A.D. 1617.

LVIII. DITTO c. A.D. 1701.

LIX. DITTO c. A.D. 1772.

LX. DITTO c. A.D. 1795.

LXI. DITTO c. A.D. 1809.

LXII. THE SPANISH KINGDOMS, A.D. 1030.

LXIII. DITTO DITTO A.D. 1210.

LXIV. DITTO DITTO A.D. 1360.

LXV. THE SPANISH KINGDOMS, AND THEIR EUROPEAN DEPENDENCIES UNDER CHARLES THE FIFTH.

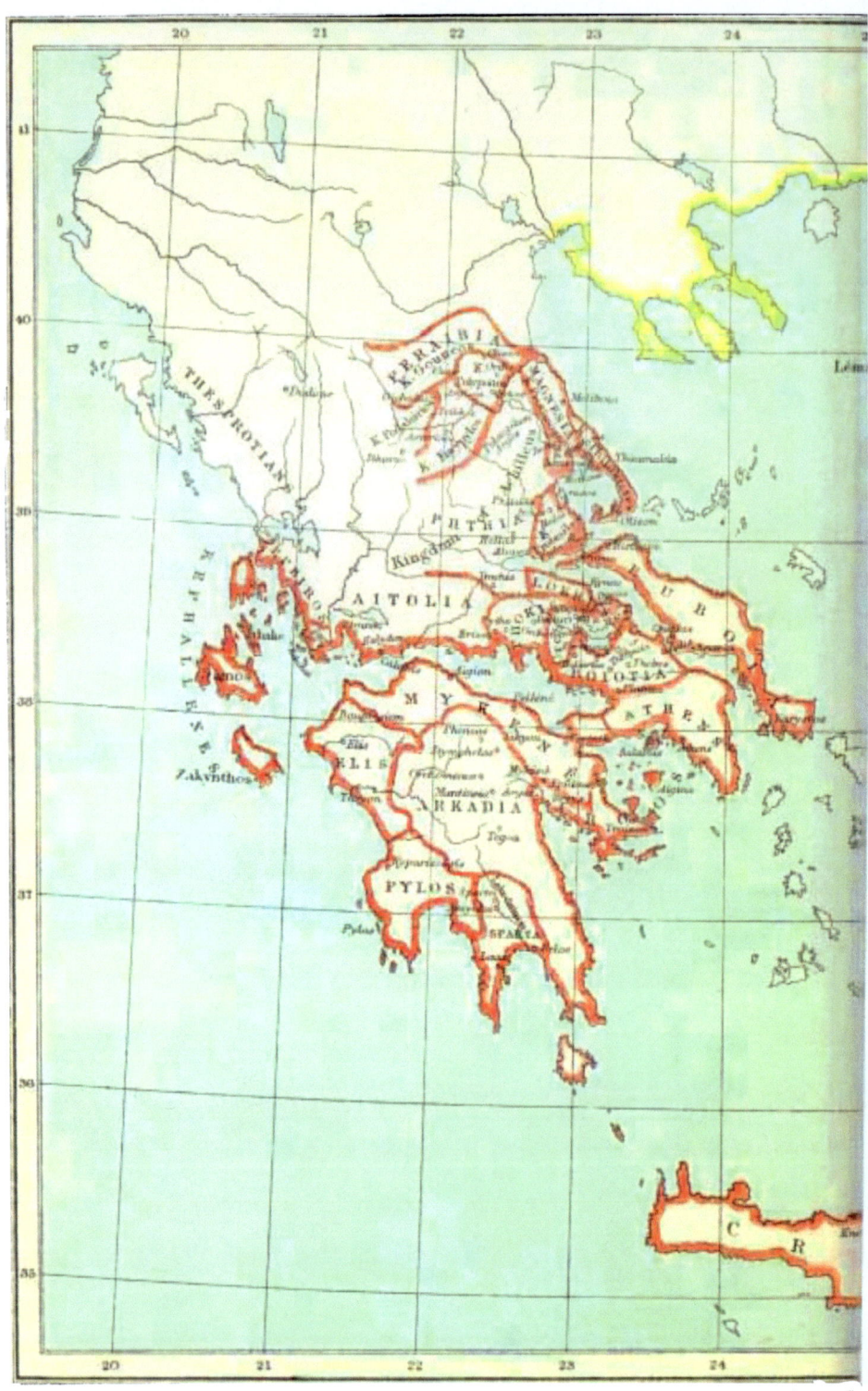

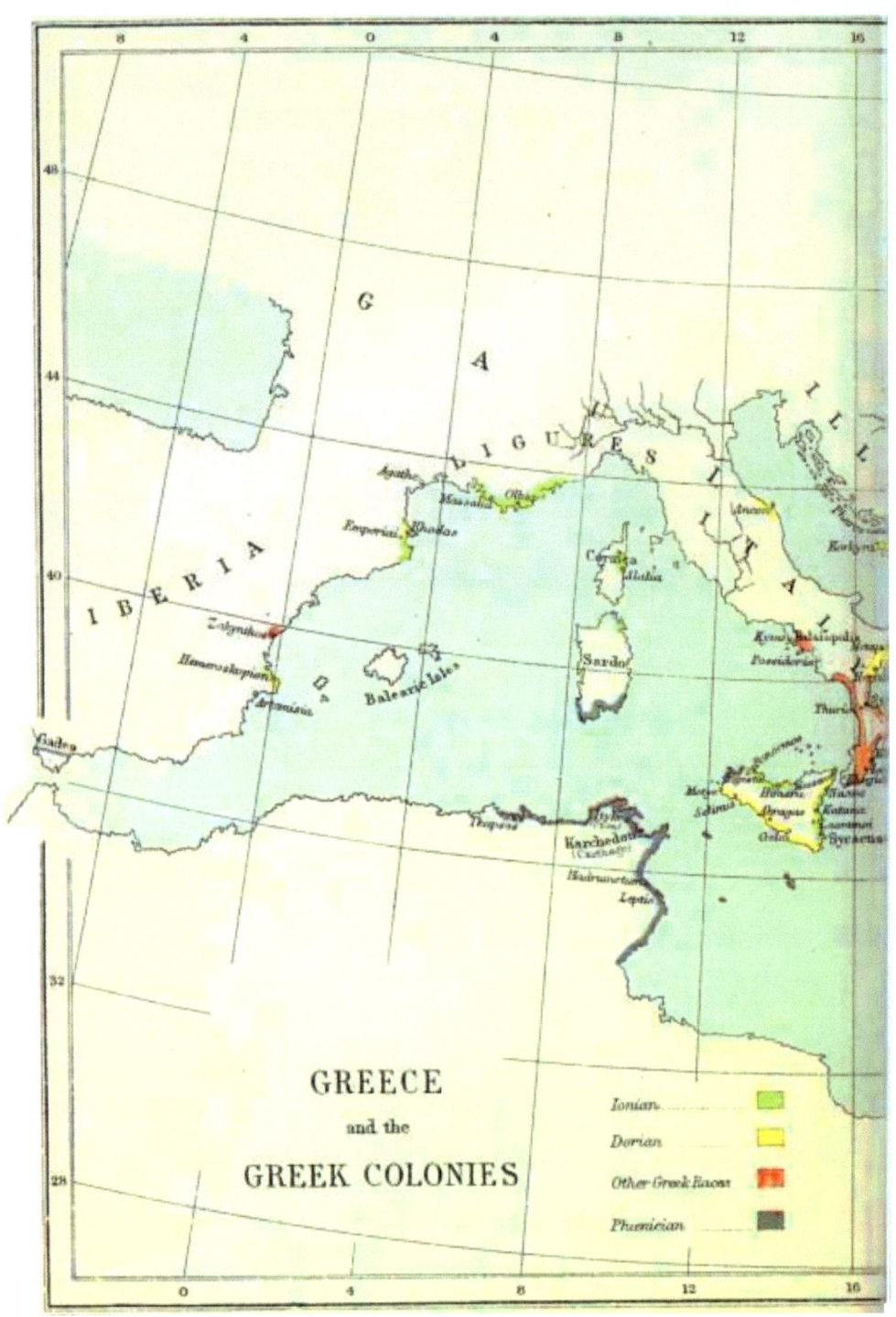

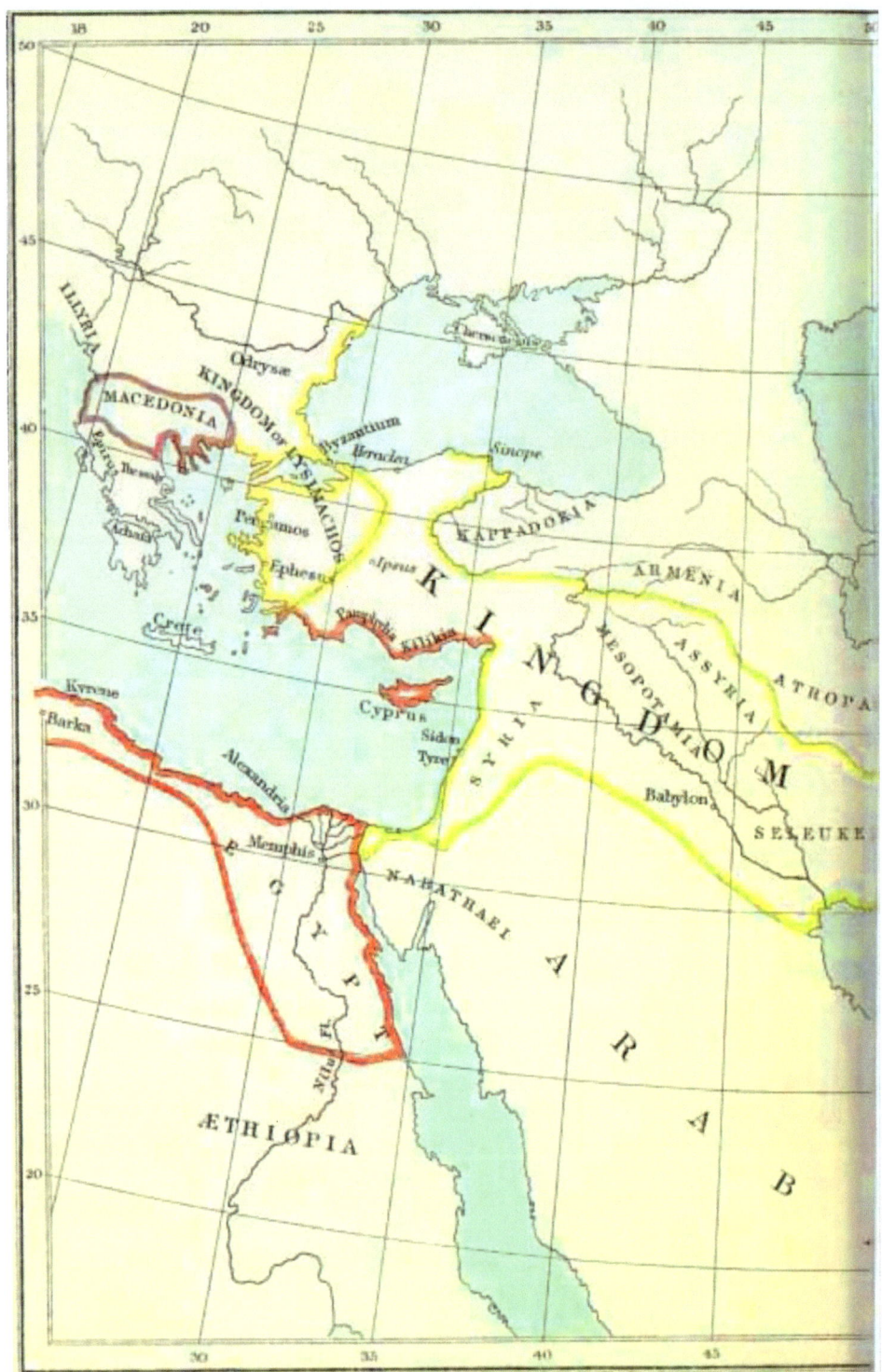

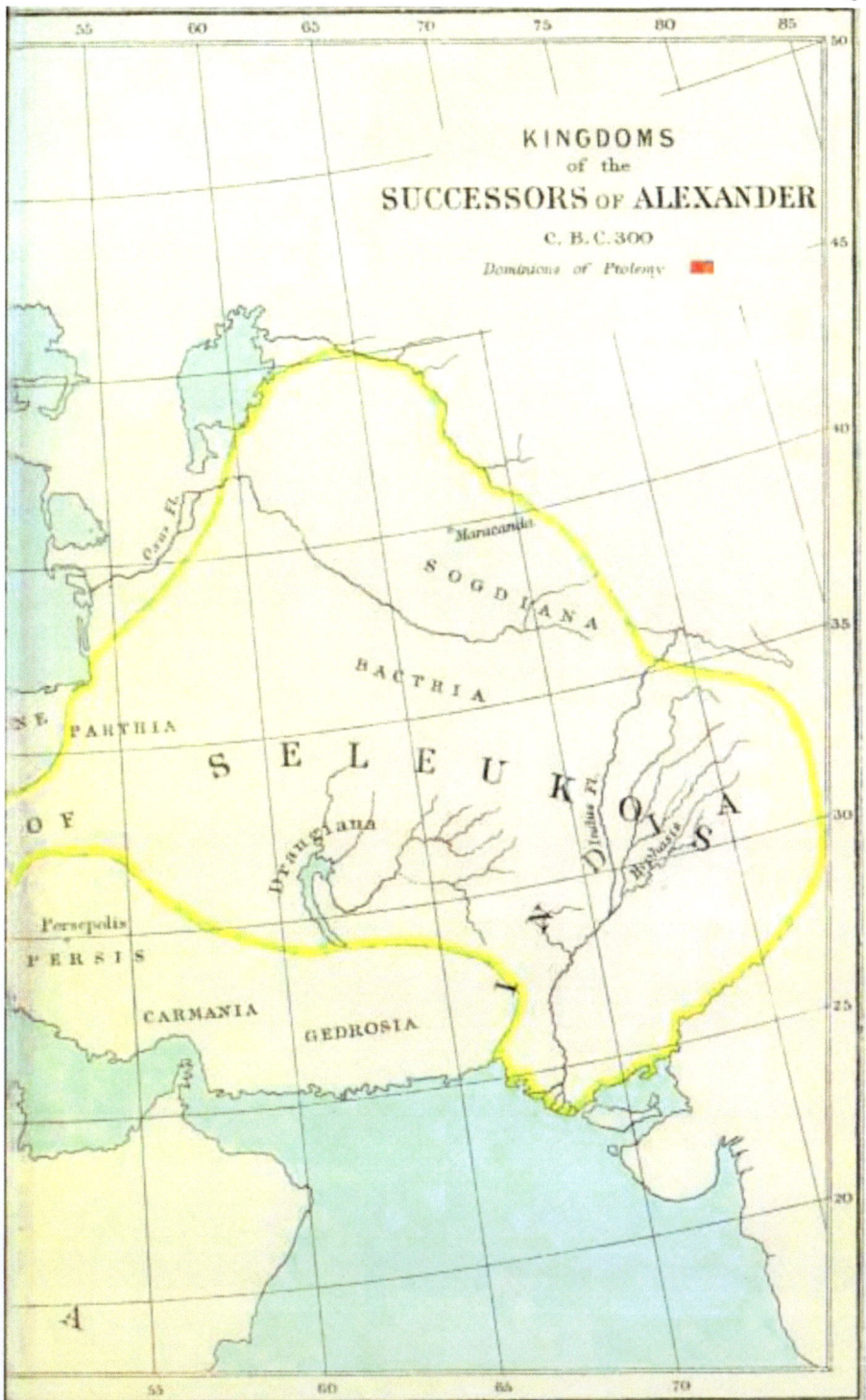

VII

Ligurians

G a u l s

Venetia

Umbrians
Etruria
Sabines
Æqui
Latins
Volscians
Samnites
Apulians
Campania

Roma
Veio
Tarquinii

CORSICA

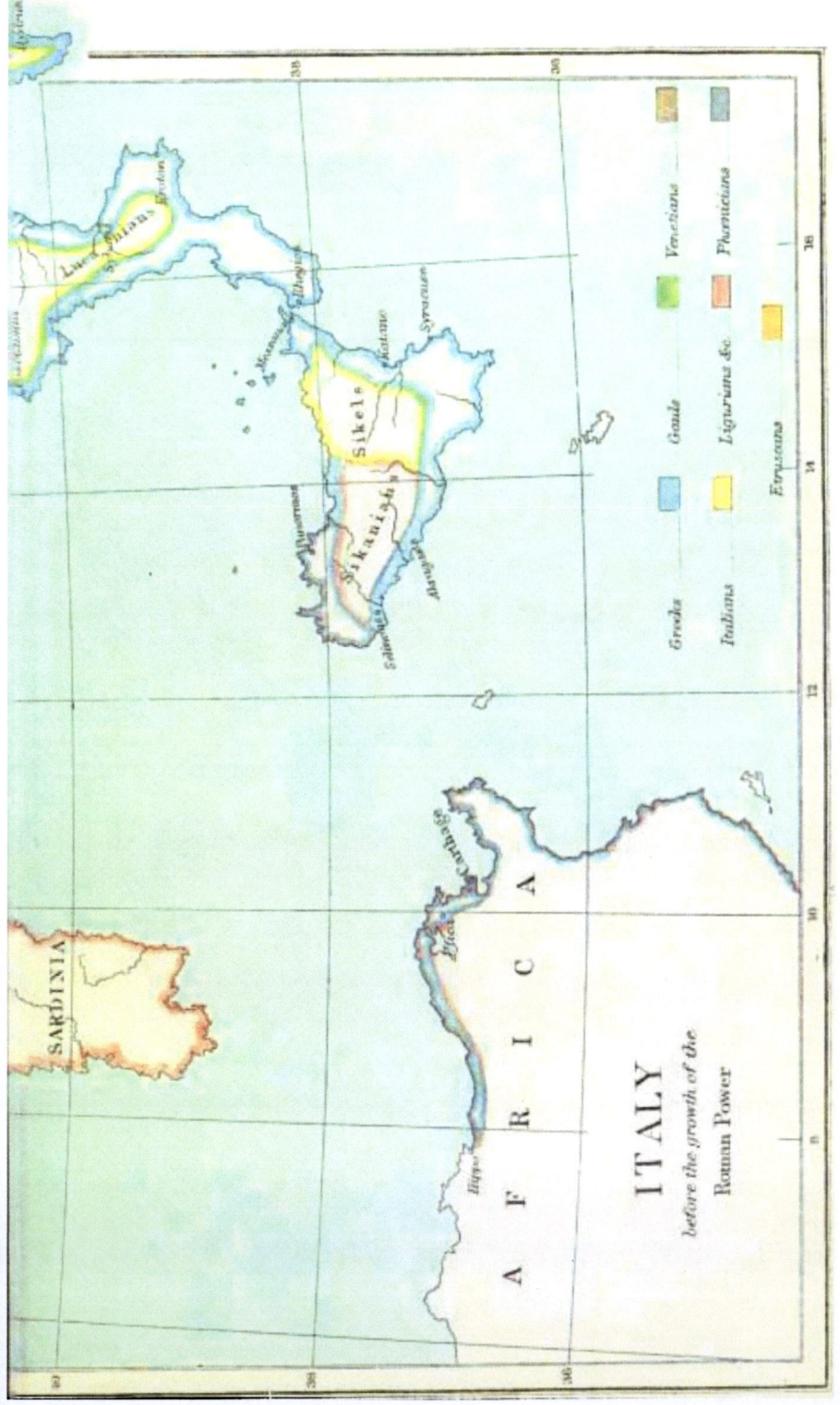

THE ROMAN DOMINIONS
at the end of the
MITHRIDATIC WAR
B.C. 64

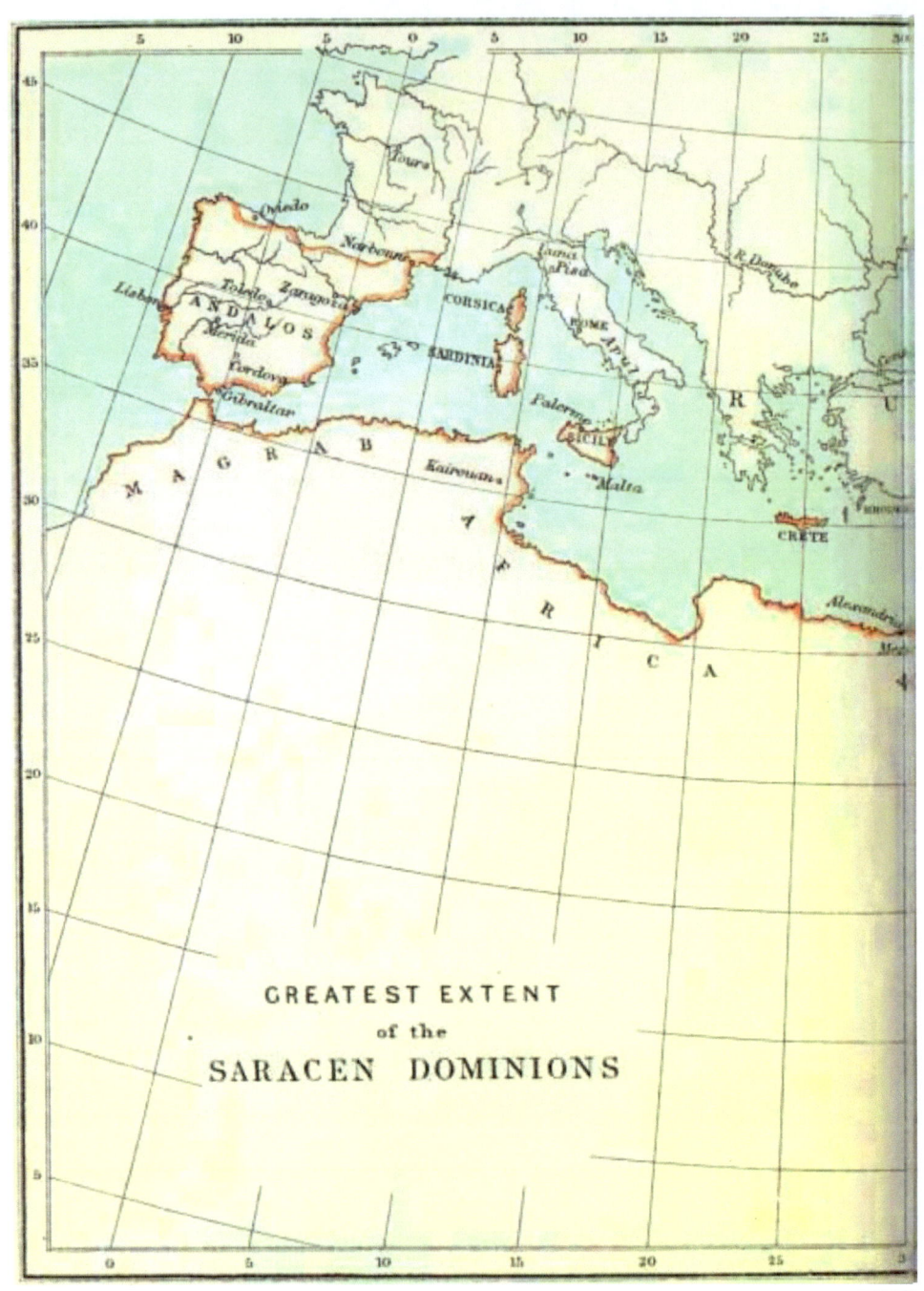

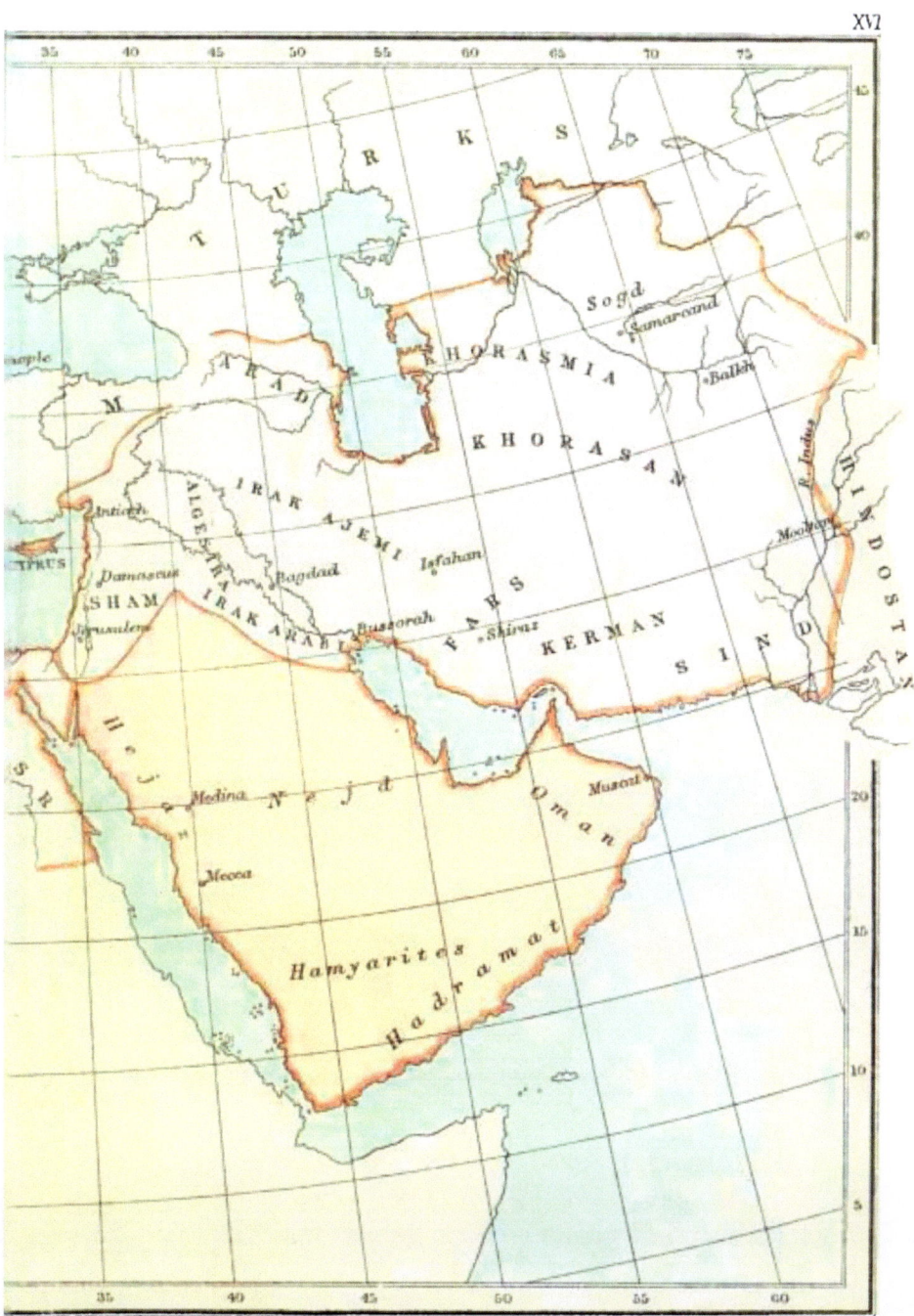

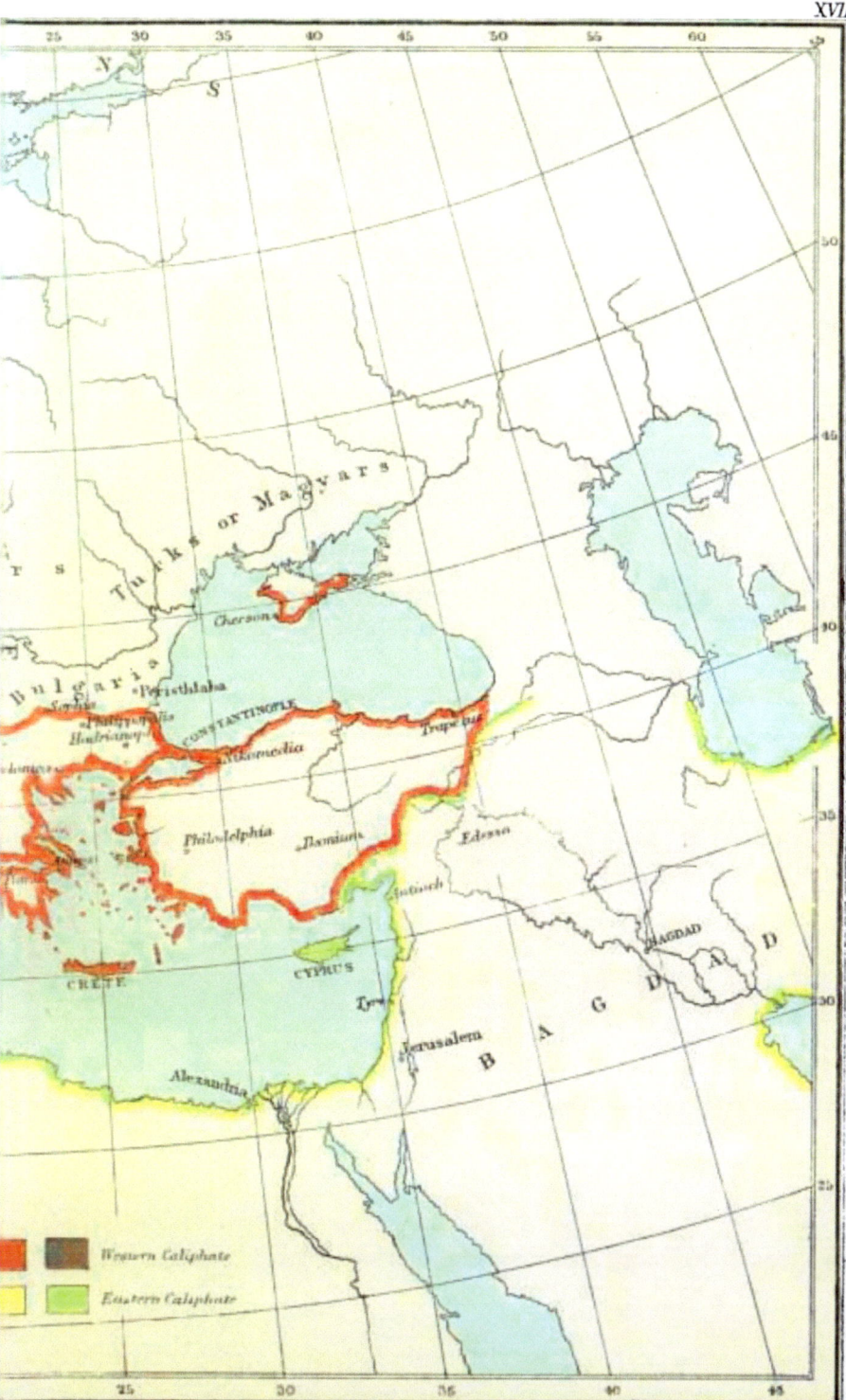

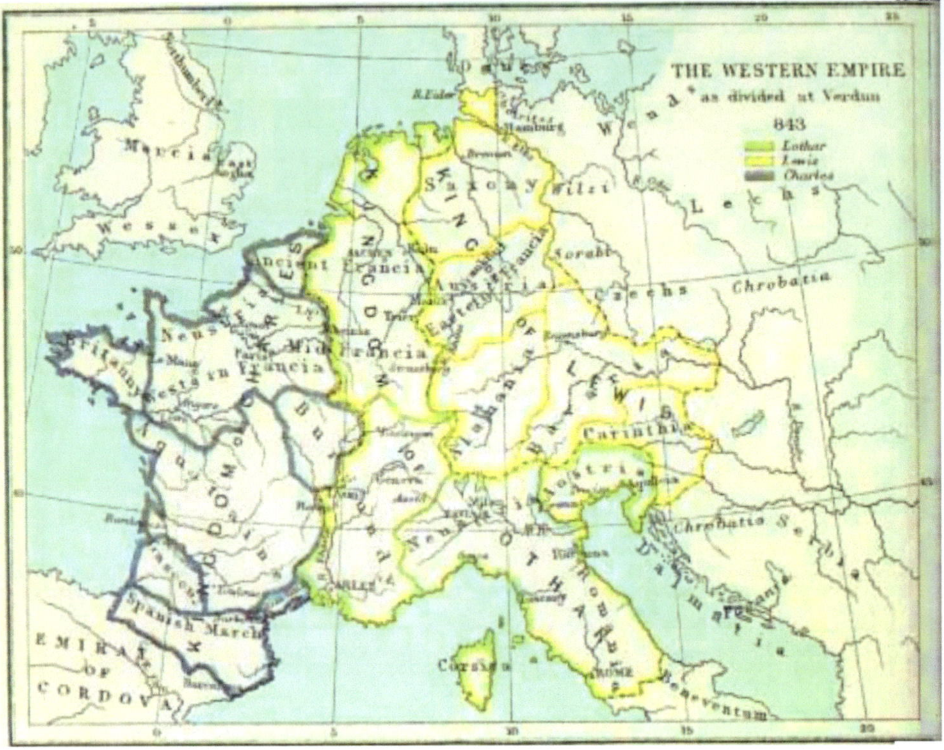
THE WESTERN EMPIRE
as divided at Verdun
843
Lothar
Lewis
Charles
Mercia
Wessex
Saxony
Ancient Francia
Austrasia
Neustria
Mid Francia
Western Francia
Carinthia
Czechs
Chrobatia
Sorabi
Lechs
Spanish March
EMIRATE OF CORDOVA
Corsica
ROME
Beneventum

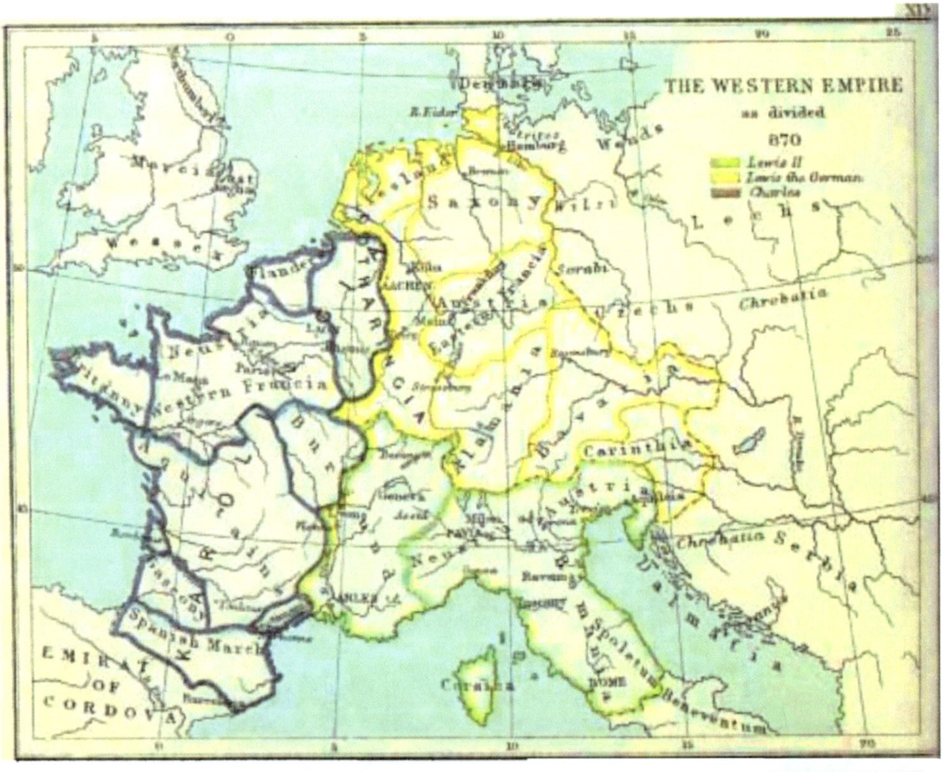
THE WESTERN EMPIRE
as divided
870
Lewis II
Lewis the German
Charles
Denmark
Mercia
Wessex
Saxony
Flanders
Austrasia
Neustria
Western Francia
Carinthia
Austria
Czechs
Chrobatia
Lechs
Spanish March
EMIRATE OF CORDOVA
Corsica
ROME
Spoletum
Beneventum

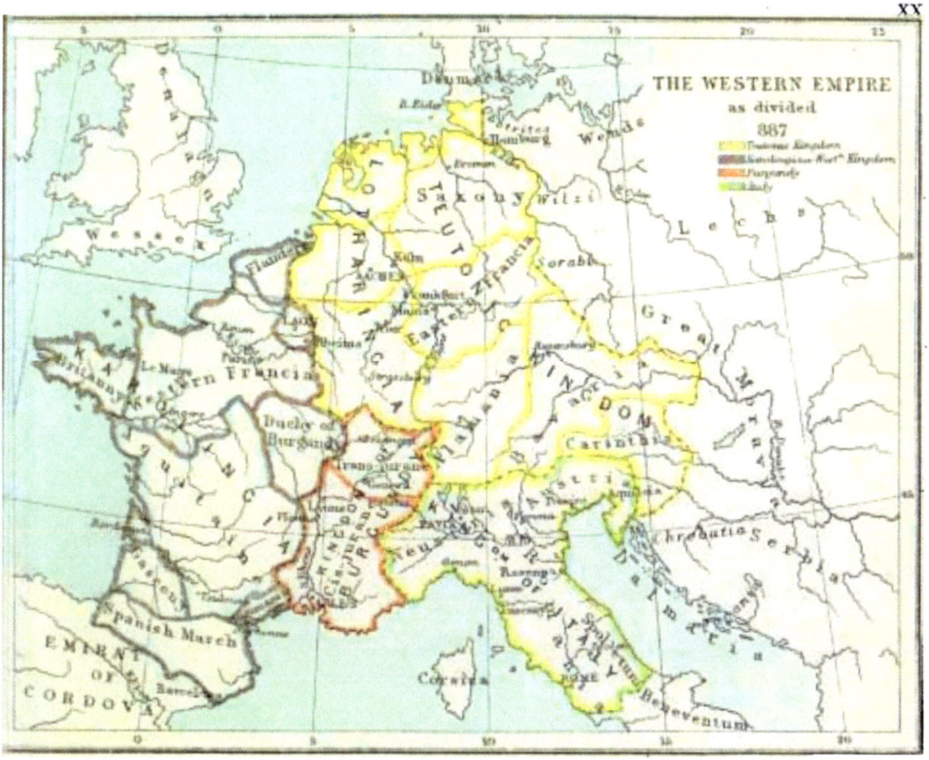

THE WESTERN EMPIRE
as divided
887

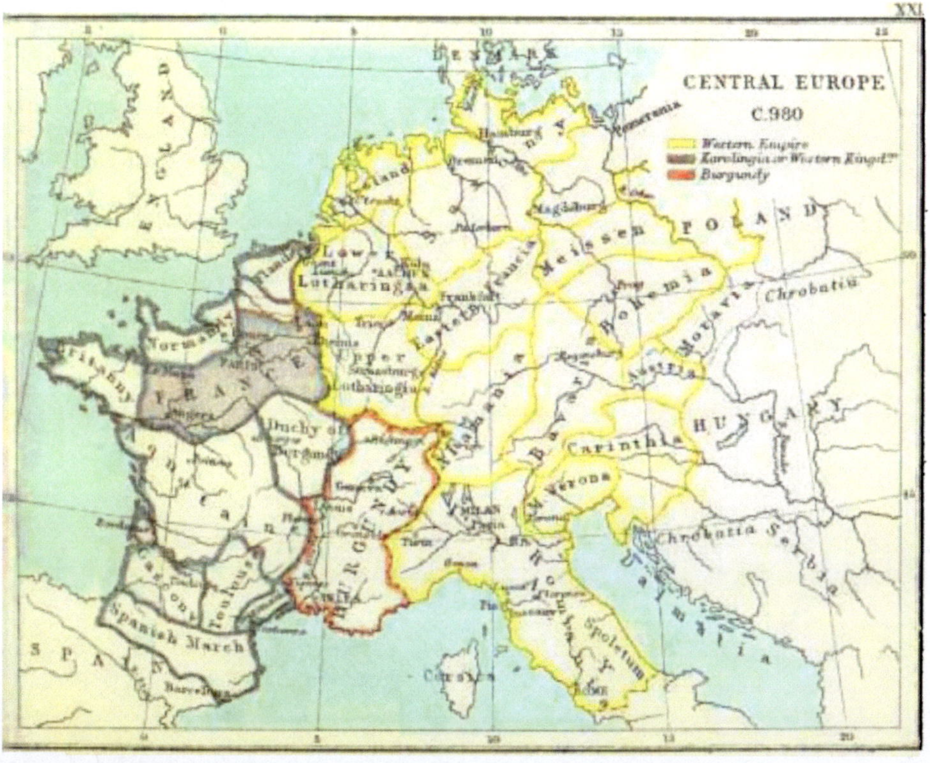

CENTRAL EUROPE
C. 980

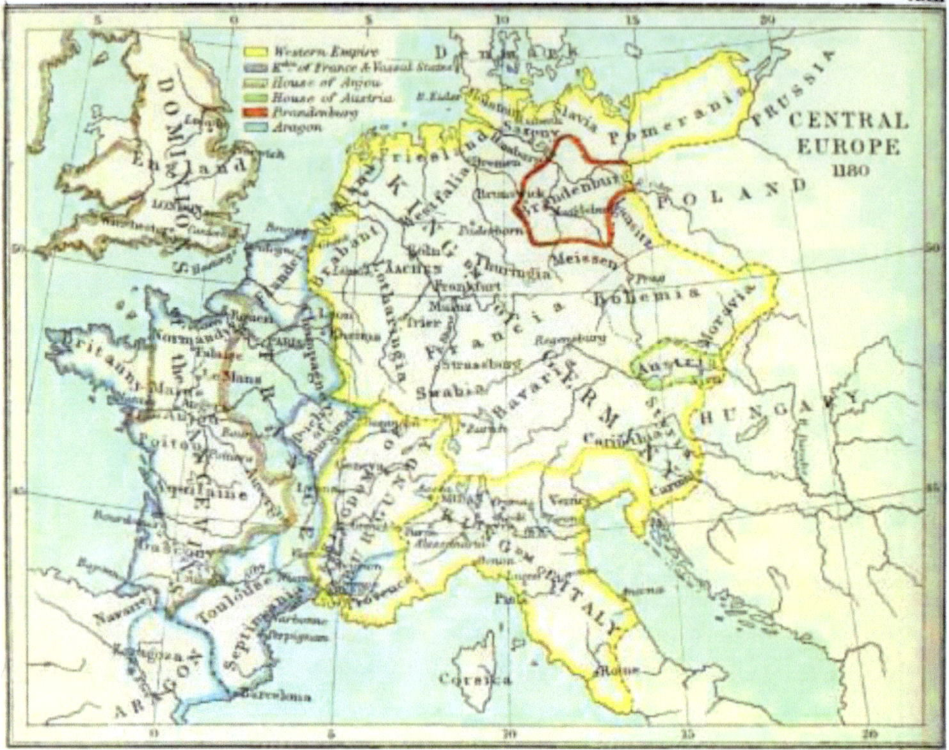
CENTRAL
EUROPE
1180
Western Empire
Kdm. of France & Vassal States
House of Anjou
House of Austria
Brandenburg
Aragon
PRUSSIA
POLAND
HUNGARY
ITALY
Corsica
Brandenburg
Bohemia
Austria
Pomerania
Francia
Bavaria
Swabia
Thuringia
Westfalia
Friesland
Brabant
Flanders
Normandy
Poitou
Aquitaine
Toulouse
Septimania
Navarre
Barcelona
England

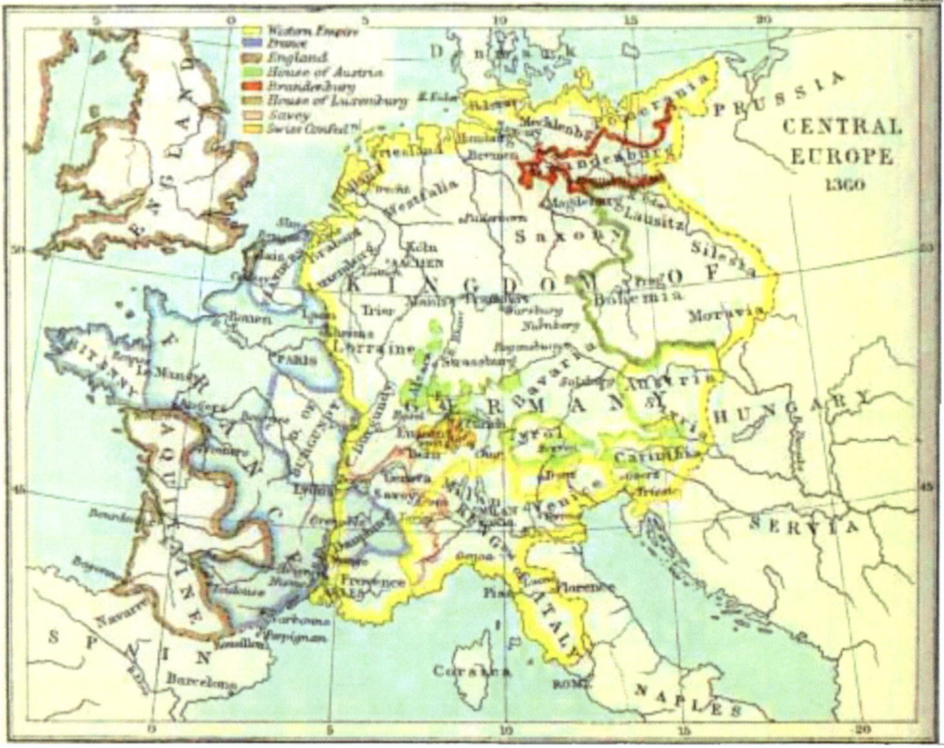
CENTRAL
EUROPE
1360
Western Empire
France
England
House of Austria
Brandenburg
House of Luxemburg
Savoy
Swiss Confedn.
PRUSSIA
HUNGARY
SERVIA
NAPLES
SPAIN
ENGLAND
KINGDOM OF GERMANY
Brandenburg
Pomerania
Mecklenbg
Saxony
Silesia
Bohemia
Moravia
Austria
Bavaria
Tyrol
Carinthia
Lorraine
Westfalia
Friesland
Brabant
Savoy
Provence
Toulouse
Navarre
Barcelona
Corsica
Florence
Venice
Trieste

XXIV

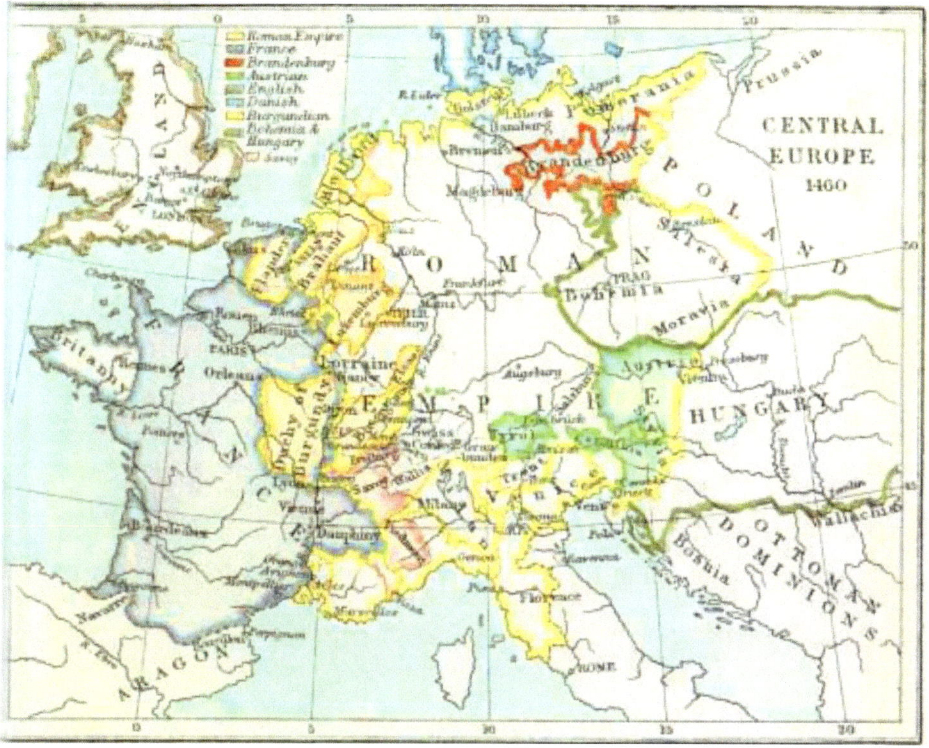
CENTRAL
EUROPE
1460
Roman Empire
France
Brandenburg
Austrian
English
Danish
Burgundian
Bohemia & Hungary
Savoy
ENGLAND
FRANCE
ROMAN EMPIRE
POLAND
PRUSSIA
HUNGARY
OTTOMAN DOMINIONS
ARAGON
Pomerania
Brandenburg
Silesia
Bohemia
Moravia
Austria
Britanny
Orleans
PARIS
Lorraine
Duchy of Burgundy
Tyrol
Venice
Milan
Genoa
Florence
ROME
Bosnia
Wallachia
Flanders
Brabant
Lübeck
Magdeburg
Frankfurt
Mainz
Köln
PRAG
Dauphiny

XXV

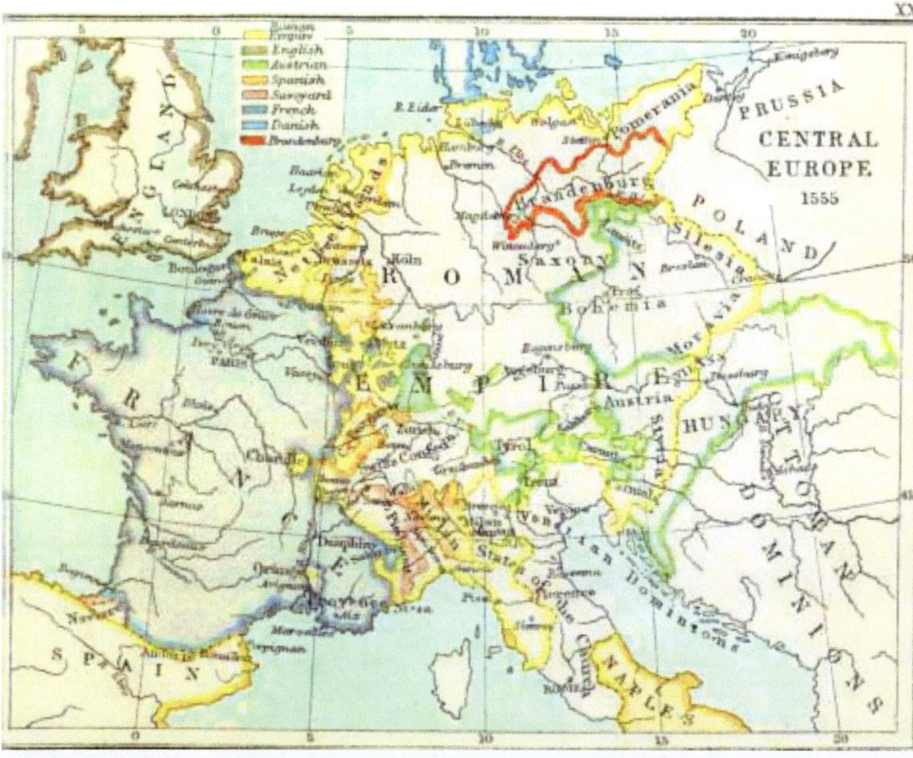
CENTRAL
EUROPE
1555
Roman Empire
English
Austrian
Spanish
Savoyard
French
Danish
Brandenburg
ENGLAND
FRANCE
ROMAN EMPIRE
PRUSSIA
POLAND
HUNGARY
OTTOMAN DOMINIONS
SPAIN
NAPLES
Pomerania
Brandenburg
Saxony
Silesia
Bohemia
Moravia
Austria
Tyrol
Venetian Dominions
States of the Church
Dauphiny
Köln
Prag
ROME

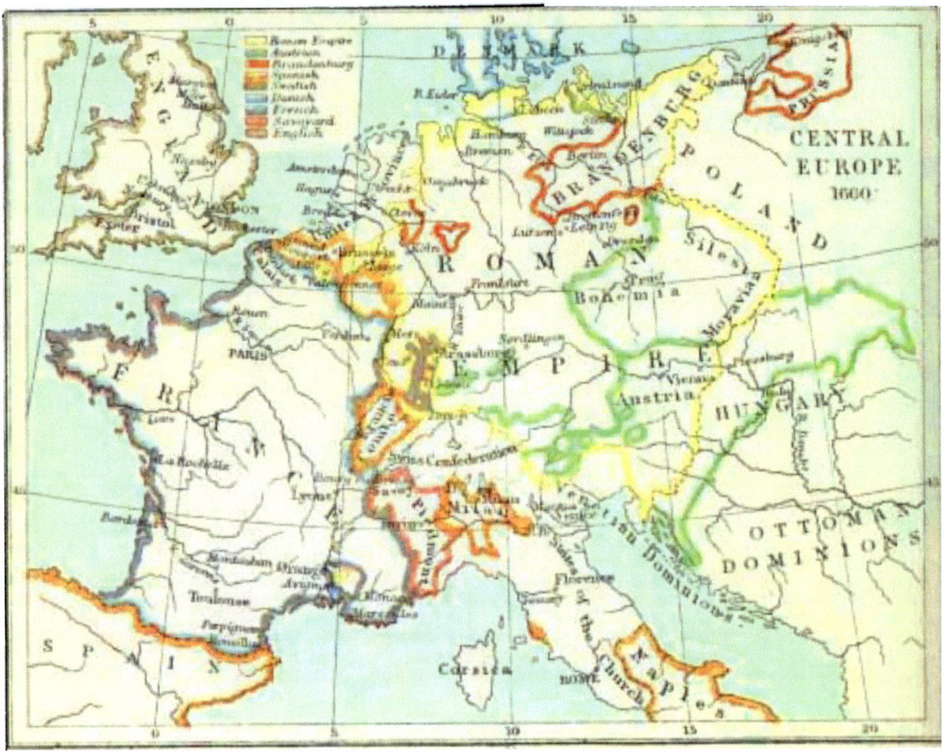
CENTRAL EUROPE 1660
Roman Empire
Austrian
Brandenburg
Spanish
Swedish
Danish
French
Savoyard
English
DENMARK
ENGLAND
BRANDENBURG
PRUSSIA
POLAND
ROMAN EMPIRE
Silesia
Bohemia
Moravia
Austria
HUNGARY
OTTOMAN DOMINIONS
Venetian Dominions
FRANCE
SPAIN
Corsica
ROME
States of the Church
Naples
Swiss Confederation
LONDON
PARIS
Berlin
Vienna
Lyons
Florence

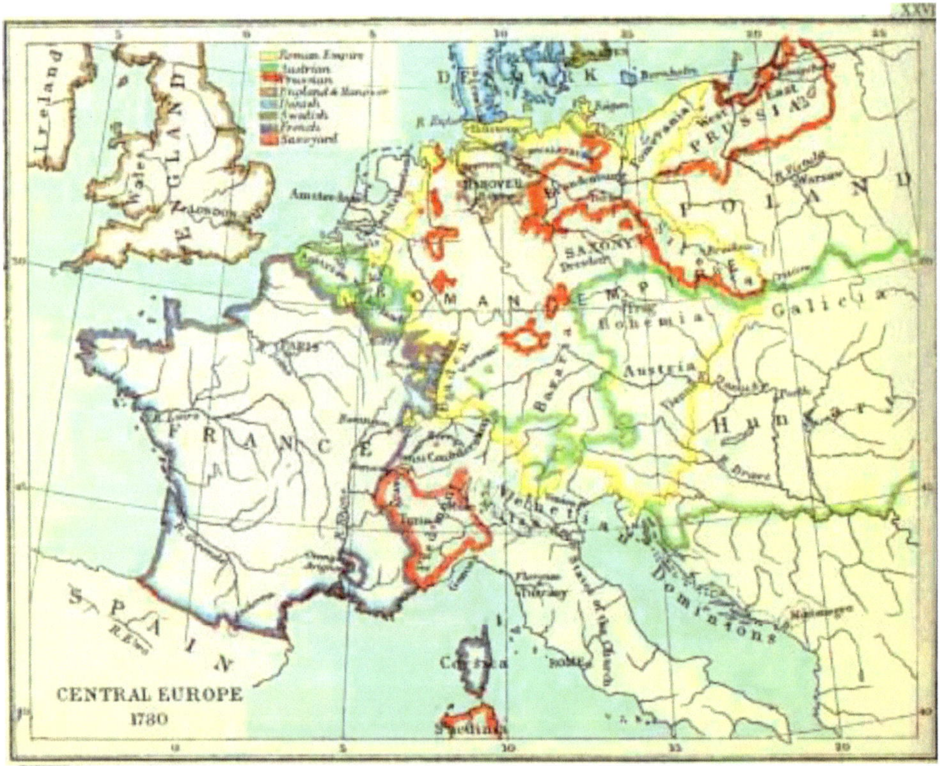
CENTRAL EUROPE 1780
Roman Empire
Austrian
Prussian
England & Hanover
Danish
Swedish
French
Savoyard
Ireland
ENGLAND
Wales
LONDON
DENMARK
PRUSSIA
POLAND
SAXONY
ROMAN EMPIRE
Galicia
Bohemia
Austria
Hungary
Bavaria
FRANCE
PARIS
SPAIN
Corsica
Sardinia
ROME
Venetian Dominions
States of the Church
Hanover

CENTRAL EUROPE
1801

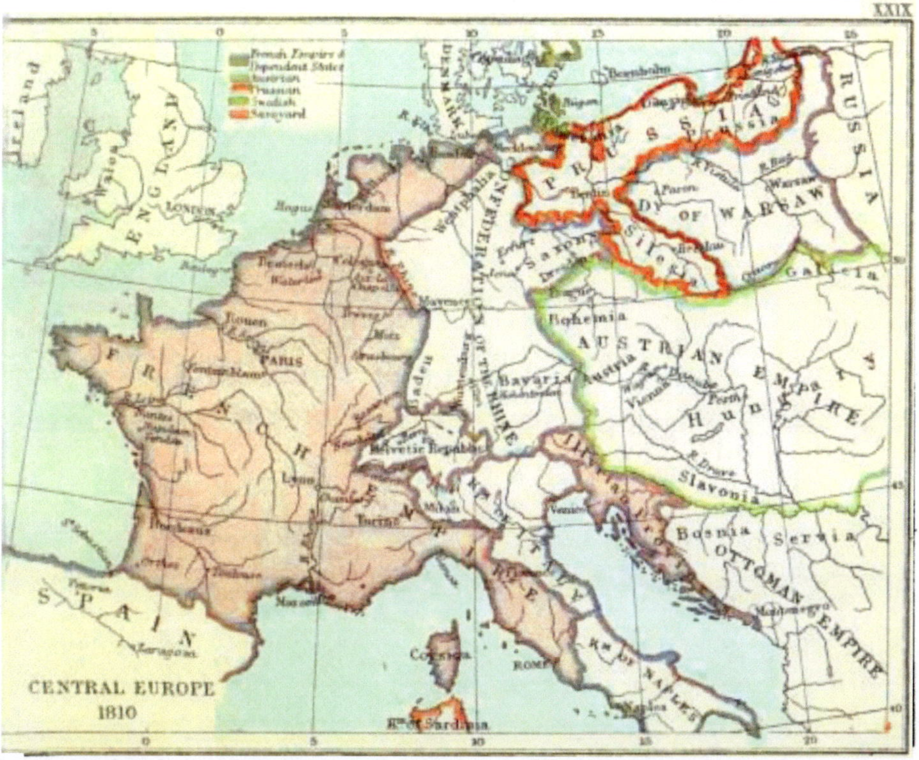

CENTRAL EUROPE
1810

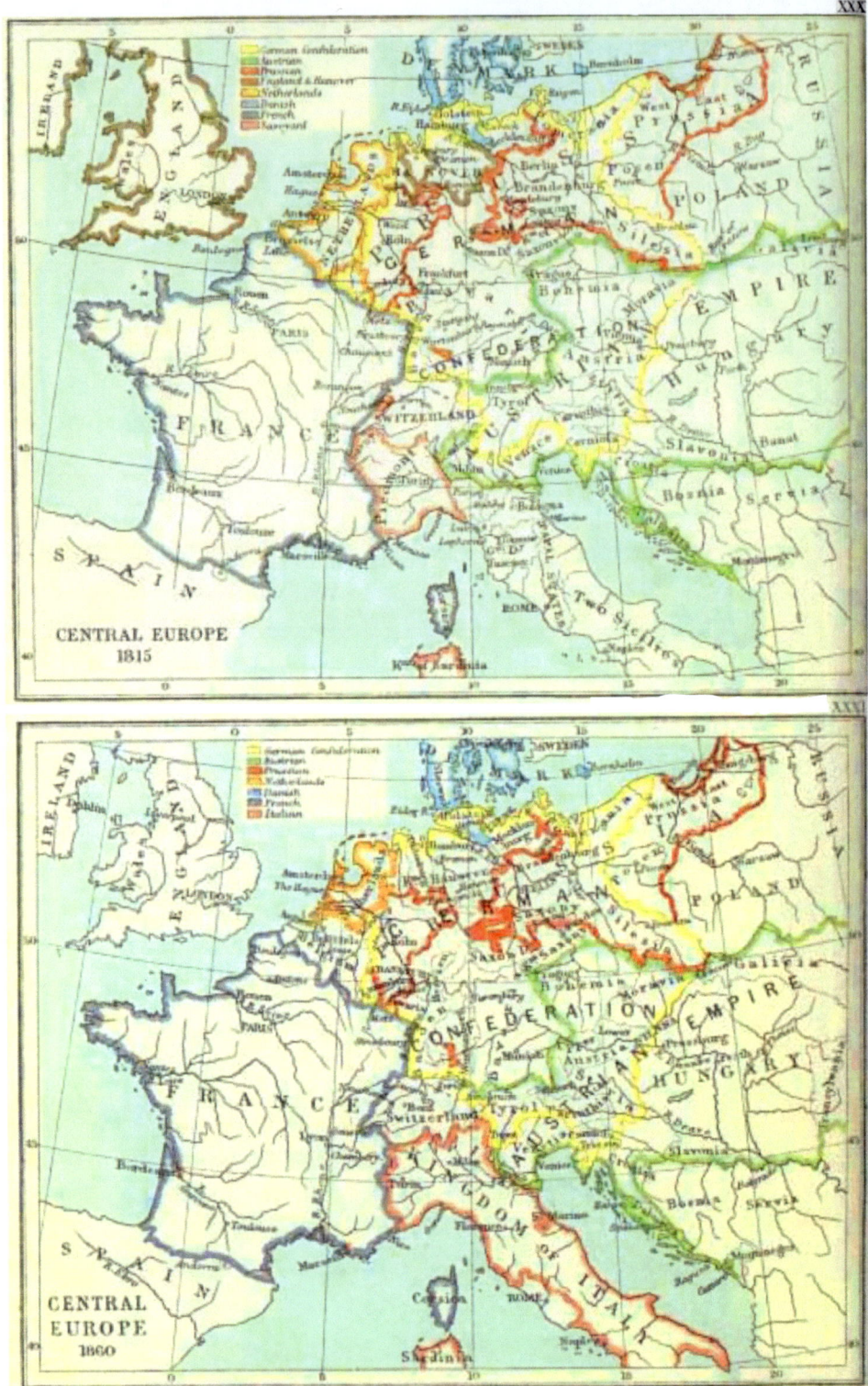

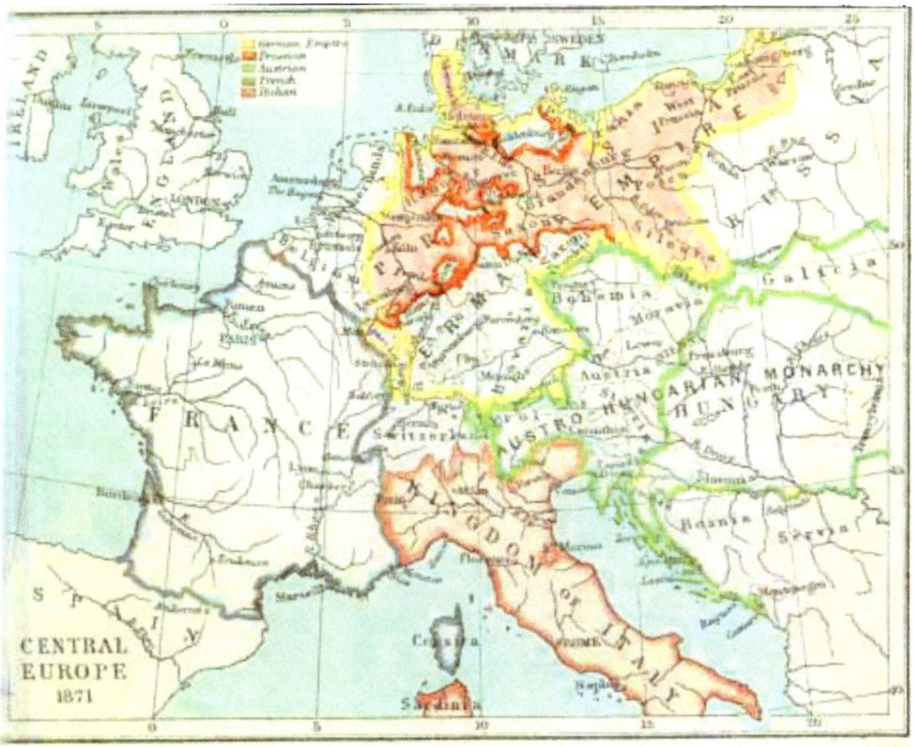

German Empire
Prussian
Austrian
French
Italian
IRELAND
ENGLAND
Wales
LONDON
Liverpool
Hull
Bristol
DENMARK
SWEDEN
Belgium
PARIS
FRANCE
Switzerland
Lyon
Bordeaux
Marseille
SPAIN
Brandenburg
Silesia
Bohemia
Moravia
Galicia
Austria
AUSTRO HUNGARIAN MONARCHY
HUNGARY
Bosnia
Servia
KINGDOM OF ITALY
ROME
Corsica
Sardinia
CENTRAL
EUROPE
1871

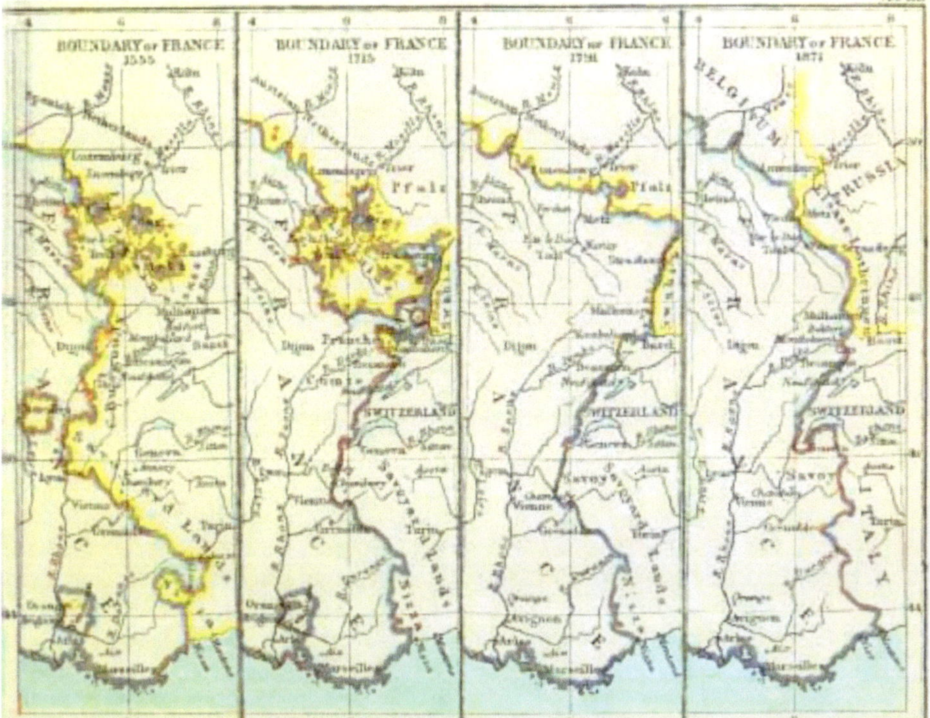

BOUNDARY OF FRANCE 1555
BOUNDARY OF FRANCE 1715
BOUNDARY OF FRANCE 1791
BOUNDARY OF FRANCE 1871
Pfalz
Dijon
SWITZERLAND
Savoy
Marseille
BELGIUM
PRUSSIA
ITALY

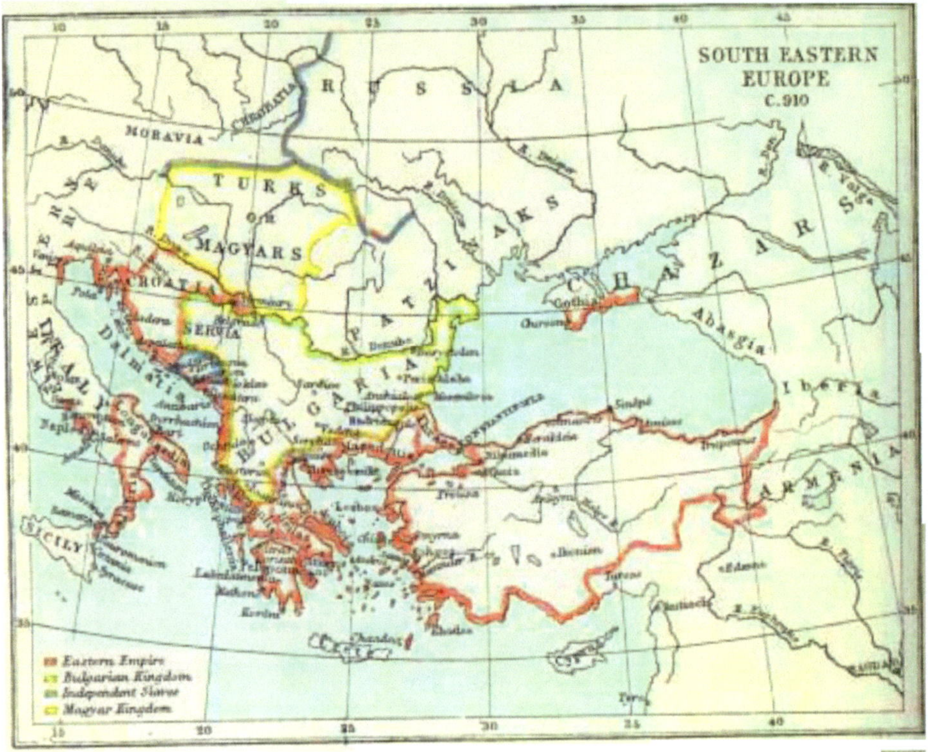
SOUTH EASTERN
EUROPE
C. 910
RUSSIA
MORAVIA
TURKS OR MAGYARS
CROATIA
SERVIA
BULGARIA
PATZINAKS
CHAZARS
Dalmatia
ITALY
SICILY
ARMENIA
Iberia
Abasgia
Constantinople
Chersonese
Gothia
Sinope
Trapezus
Rhodes
Eastern Empire
Bulgarian Kingdom
Independent Slaves
Magyar Kingdom

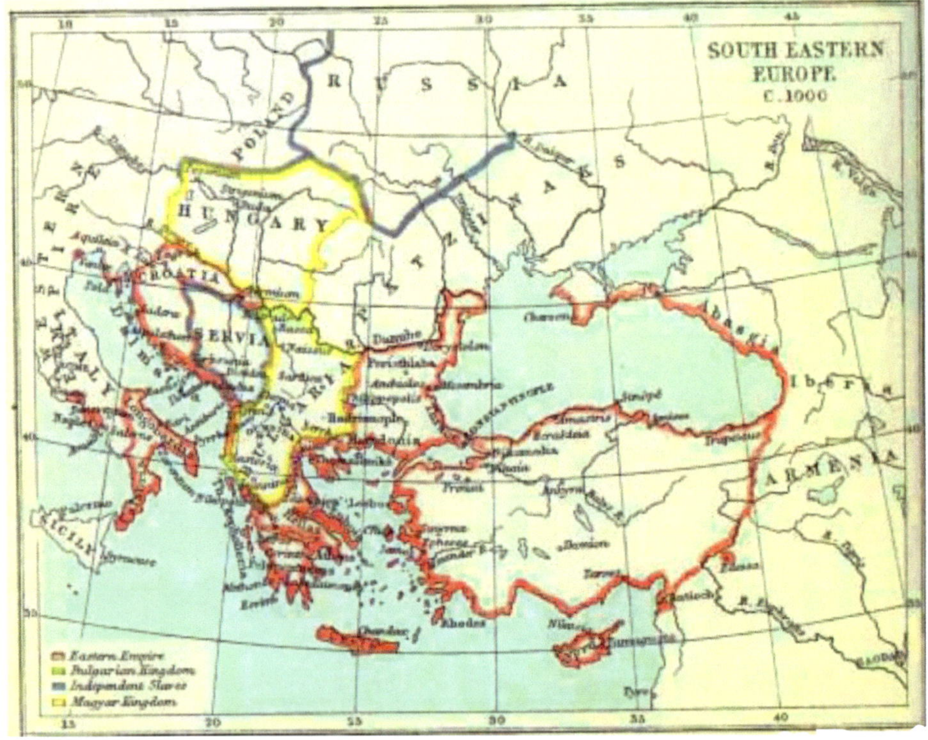
SOUTH EASTERN
EUROPE
C. 1000
RUSSIA
POLAND
HUNGARY
CROATIA
SERVIA
BULGARIA
PATZINAKS
ITALY
SICILY
ARMENIA
Iberia
Abasgia
Constantinople
Chersonese
Sinope
Trapezus
Rhodes
Eastern Empire
Bulgarian Kingdom
Independent Slaves
Magyar Kingdom

Longmans, Green & Co.

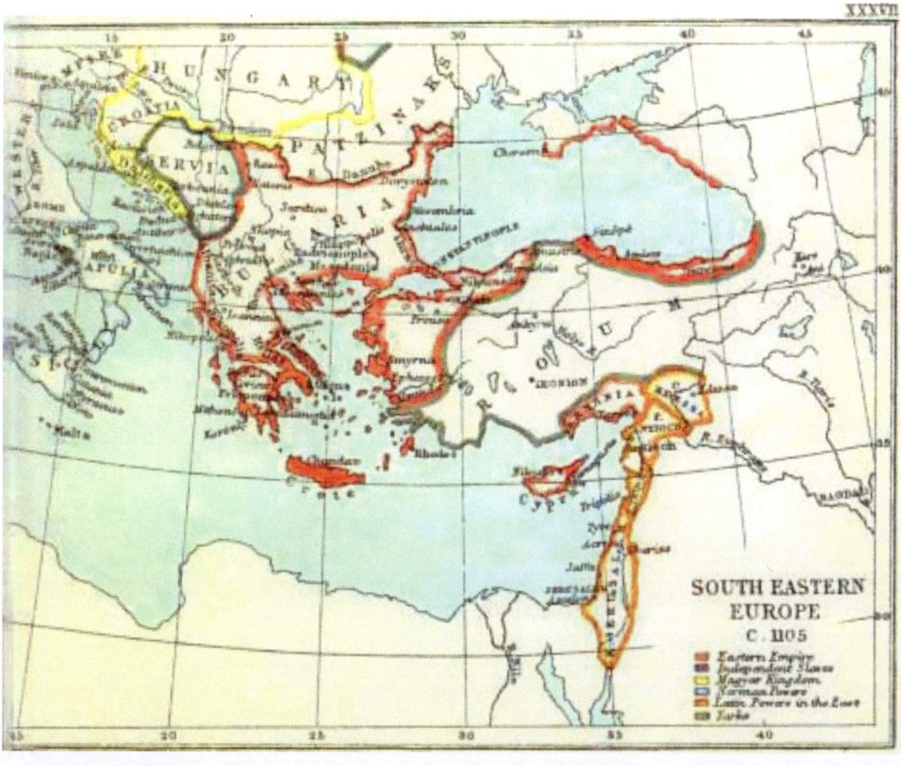

SOUTH EASTERN EUROPE C. 1040-1070

SOUTH EASTERN EUROPE C. 1105

SOUTH EASTERN EUROPE
C. 1180
POLAND
RUSSIA
HUNGARY
WESTERN EMPIRE
CROATIA
SERVIA
CUMANS
BULGARIA
KINGDOM OF SICILY
SULTANATE OF ROUM
ARMENIA
Transylvania
R. Danube
Eastern Empire
Magyar Kingdom
Latin Powers in the East

SOUTH EASTERN EUROPE
C. 1210
POLAND
RUSSIA
Bohemia
Galicia
HUNGARY
WESTERN EMPIRE
Transylvania
Hermanstadt
K. BULGARIA
KINGDOM OF SICILY
EMPIRE OF NICAEA
ARMENIA
K. CYPRUS
Rhodes
Jerusalem
Latin Powers in the East
Magyar Kingdom

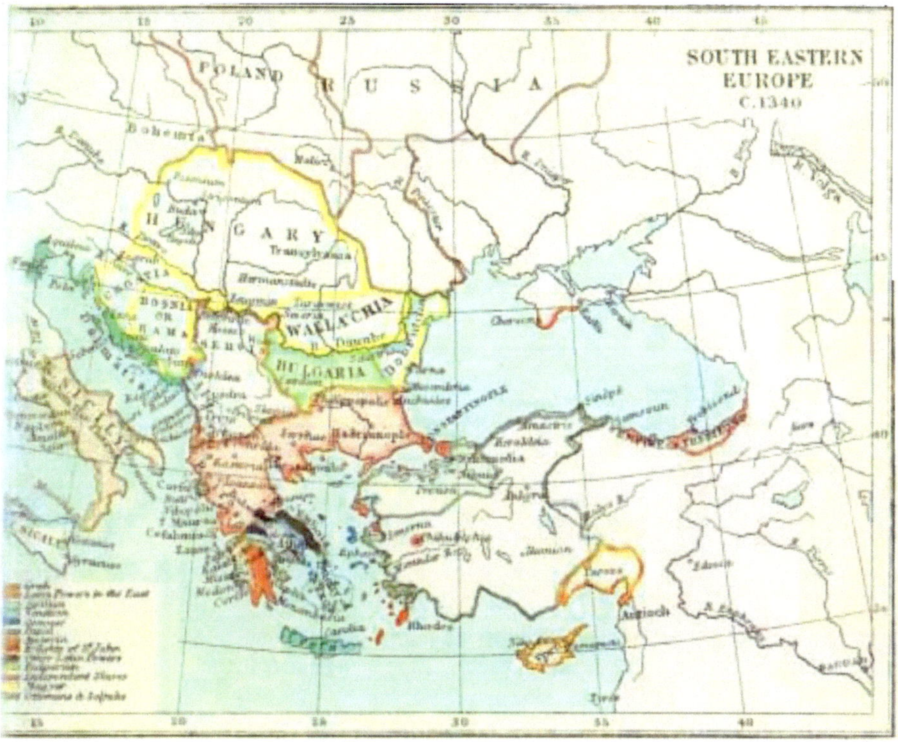

SOUTH EASTERN EUROPE
C. 1340

SOUTH EASTERN EUROPE
C. 1354-1358

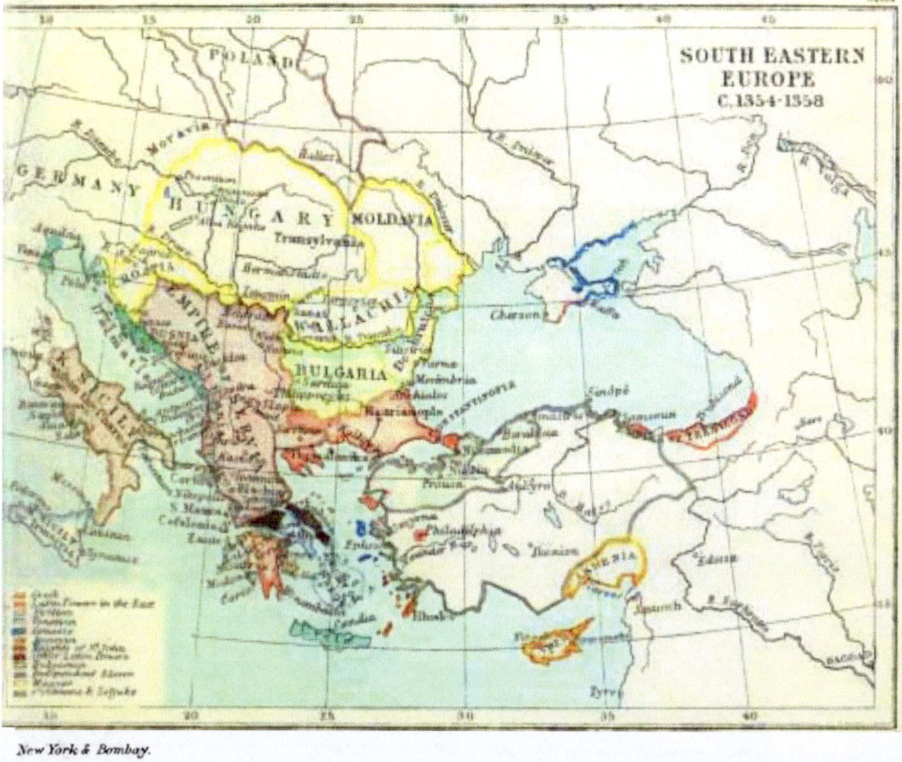

New York & Bombay.

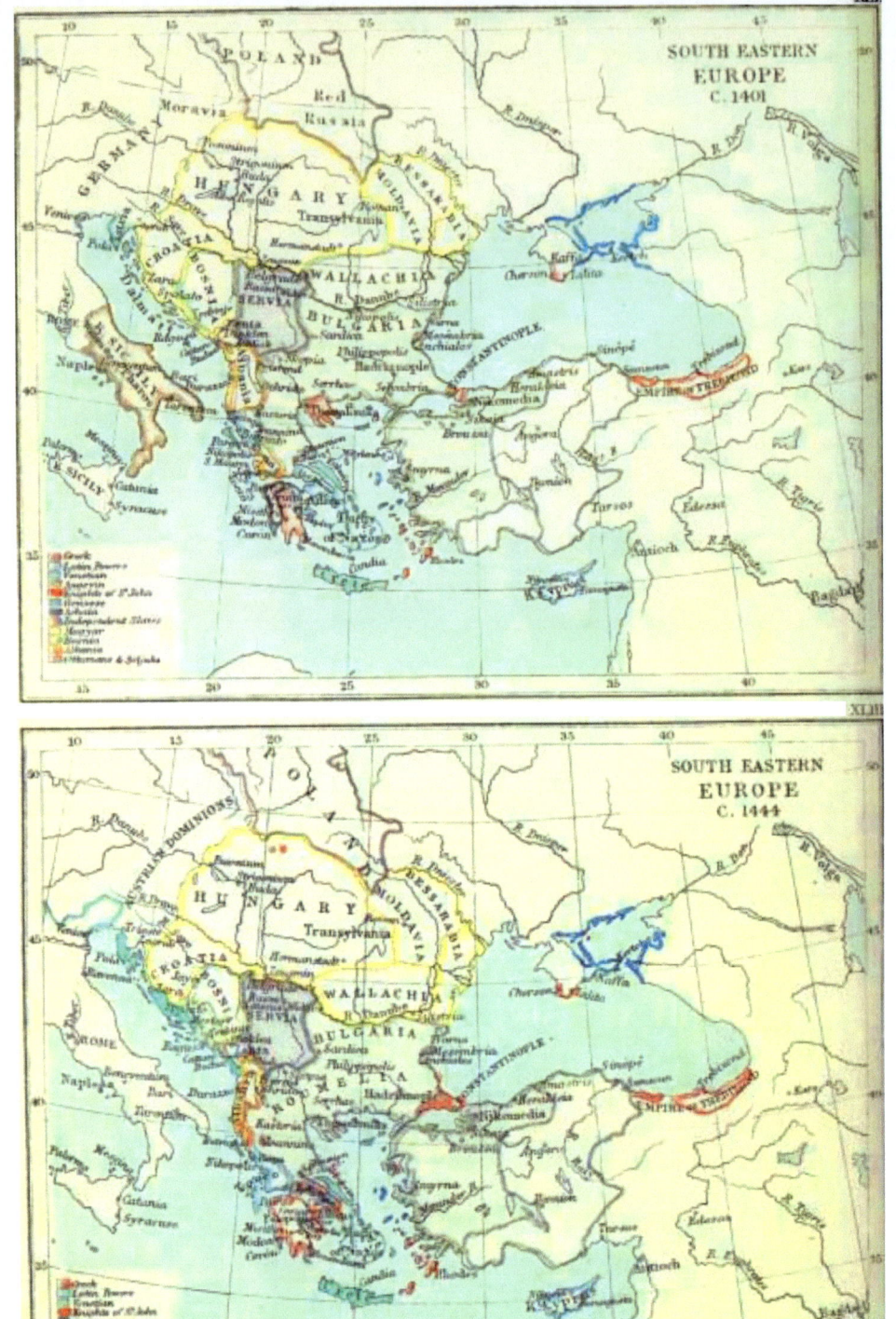

SOUTH EASTERN
EUROPE
C. 1464
HUNGARY
Transylvania
Slavonia
BOSNIA
SERVIA
WALLACHIA
BESSARABIA
BULGARIA
ROMELIA
CONSTANTINOPLE
Hadrianople
Nikomedia
Trebizond
Sinopé
Smyrna
Rhodes
Antioch
Kaffa
Rome
Naples
Bari
Catania
Syracuse
Candia
R. Danube
R. Dniester
R. Dnieper
R. Don
R. Volga
R. Tigris
R. Euphrates
Edessa
Angora
Tarsus
Silistria
Mesembria
Anchialos
Sardica
Philippolis
Kastoria
Nikopolis
D. Leukas
Broussa
Iconium
Halys R.

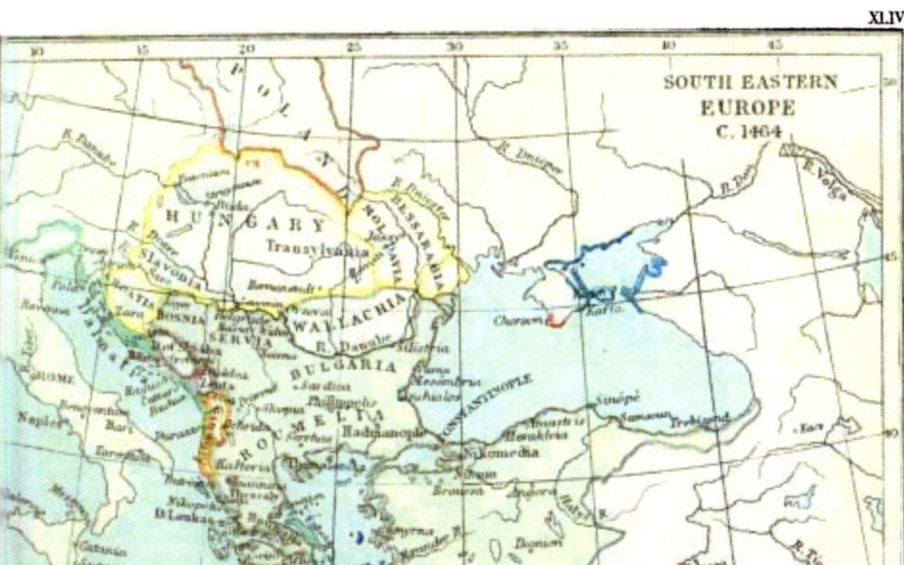

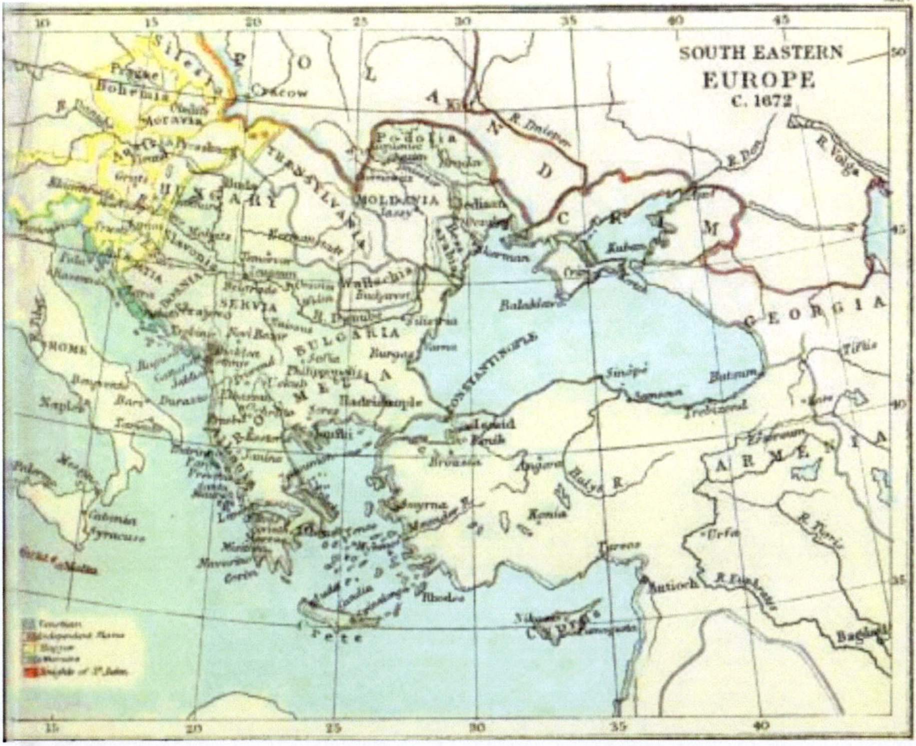
SOUTH EASTERN
EUROPE
C. 1672
POLAND
Cracow
Prague
Bohemia
Moravia
HUNGARY
TRANSYLVANIA
Podolia
MOLDAVIA
Jassy
Wallachia
Bukharest
SERVIA
Belgrade
BOSNIA
Slavonia
BULGARIA
ROMELIA
CONSTANTINOPLE
Hadrianople
Sofia
Philippopolis
Burgas
CRIM
Kuban
GEORGIA
Tiflis
Batum
Trebizond
Sinopé
ARMENIA
Erzeroum
Angora
Konia
Smyrna
Broussa
Rhodes
Antioch
Tarsus
Urfa
Cyprus
Crete
Candia
Rome
Naples
Bari
Catania
Syracuse
Malta
R. Danube
R. Dnieper
R. Don
R. Volga
R. Tigris
R. Euphrates
Halys R.
Balaklava
Ismid

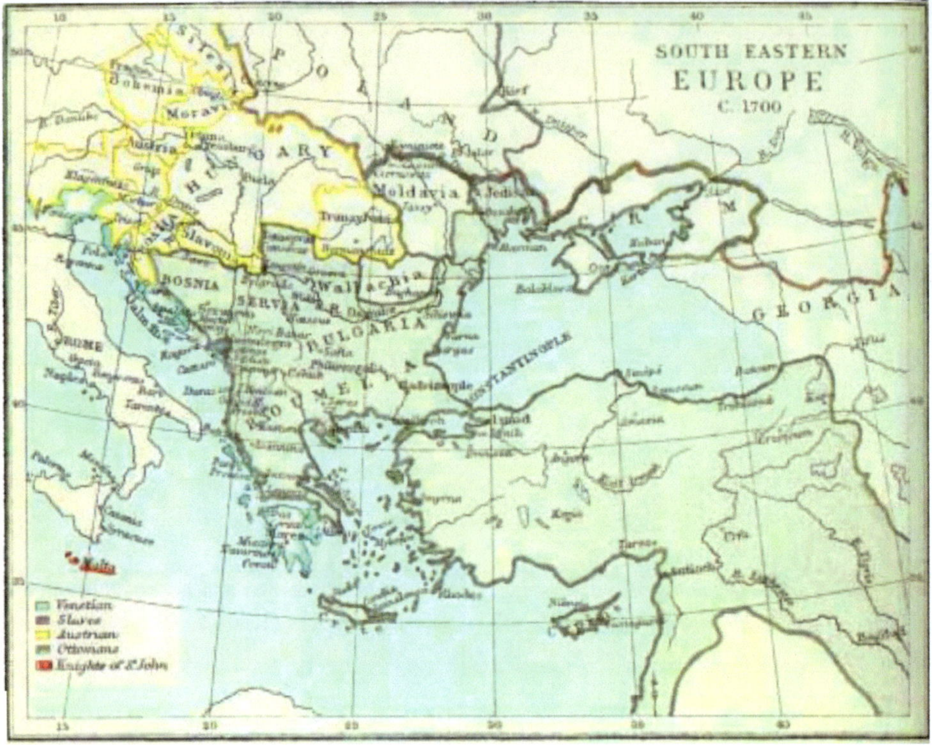
SOUTH EASTERN
EUROPE
C. 1700
POLAND
Bohemia
Austria
Moldavia
Transylvania
Wallachia
BOSNIA
SERVIA
BULGARIA
ROUMELIA
CONSTANTINOPLE
GEORGIA
ROME
Malta
Crete
Venetian
Slaves
Austrian
Ottomans
Knights of St. John

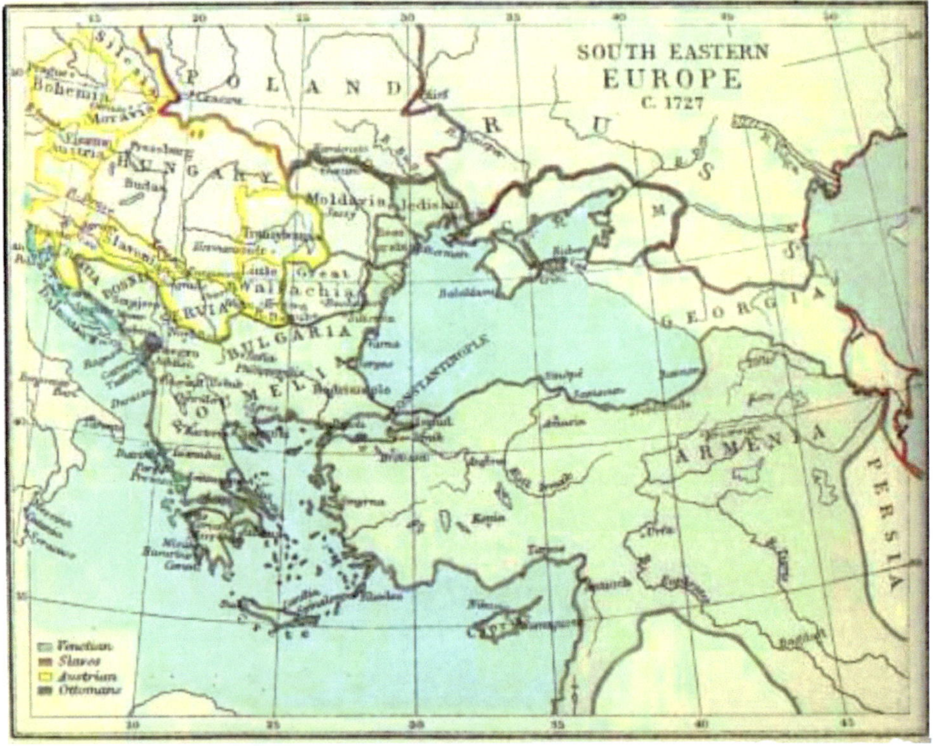
SOUTH EASTERN
EUROPE
C. 1727
POLAND
RUSSIA
Bohemia
Moravia
Austria
HUNGARY
Moldavia
Jedisan
Transylvania
Little
Great
Wallachia
SERVIA
BULGARIA
ROUMELIA
CONSTANTINOPLE
GEORGIA
ARMENIA
PERSIA
Crete
Venetian
Slaves
Austrian
Ottomans

XLVIII

SOUTH EASTERN EUROPE 1861

- Slaves
- Austrian
- Greek
- Roumanian
- Ottomans
- English

XLIX

SOUTH EASTERN EUROPE 1881

- Slaves
- Austrian
- Greek
- Roumanian
- Bulgarian
- Ottomans
- English

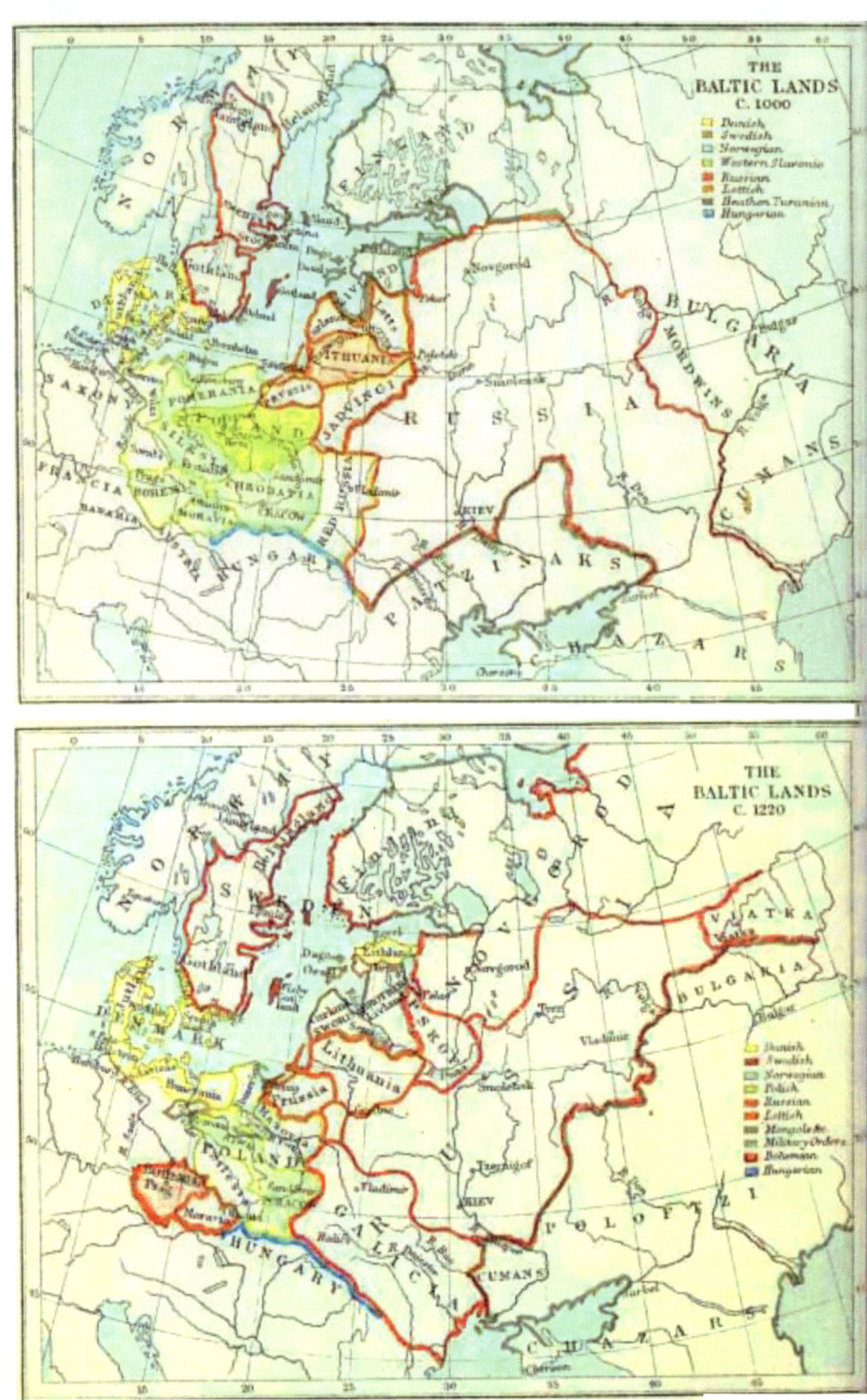

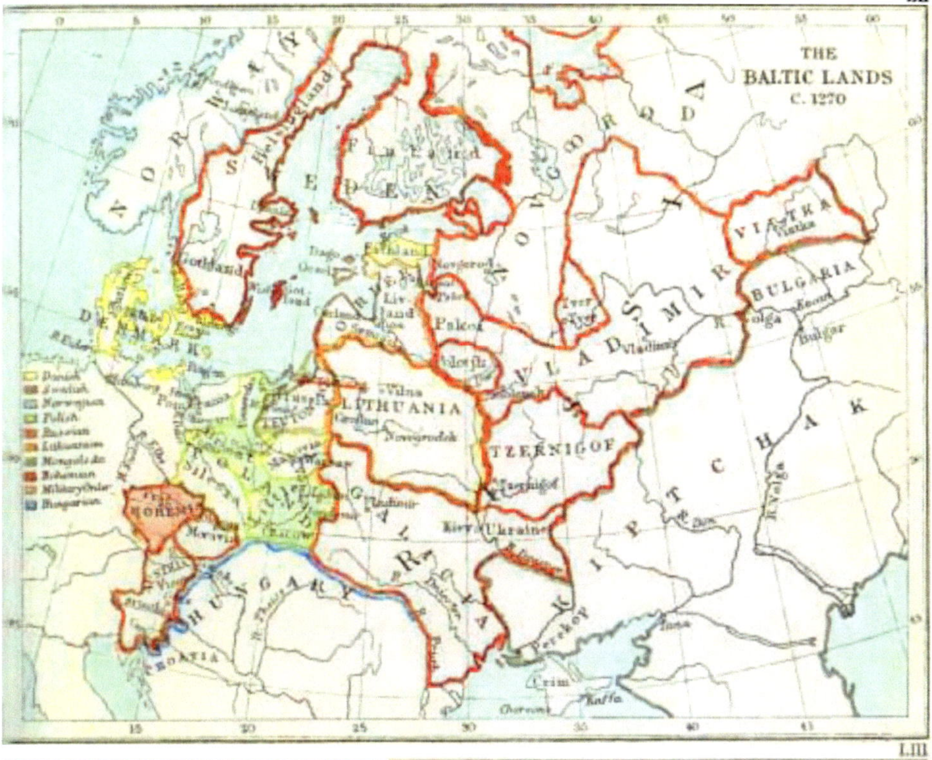

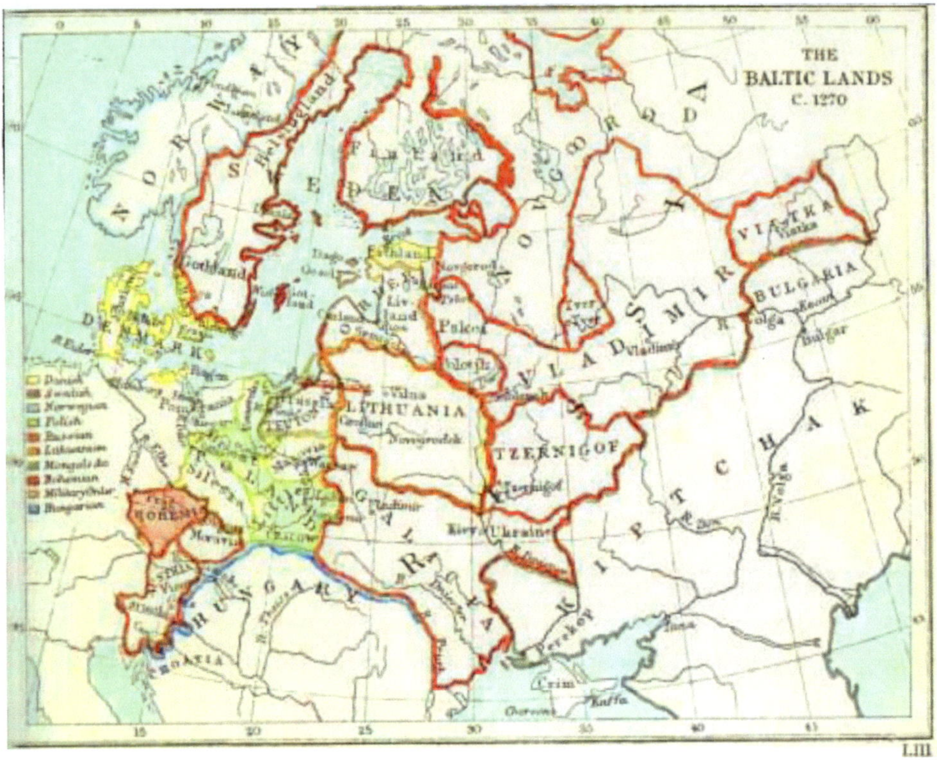
THE
BALTIC LANDS
C. 1270
Danish
Swedish
Norwegian
Polish
Russian
Lithuanian
Mongols &c.
Bohemian
Military Order
Hungarian
NORWAY
SWEDEN
NOVGOROD
VLADIMIR
VIATKA
BULGARIA
LITHUANIA
TZERNIGOF
KIPTCHAK
HUNGARY
CROATIA
DENMARK
Gothland
Novgorod
Pskof
Vilna
Kiev
Ukraine
Perekop
Crim
Kaffa
Tana
Bulgar
Cracow
Moravia

LIII

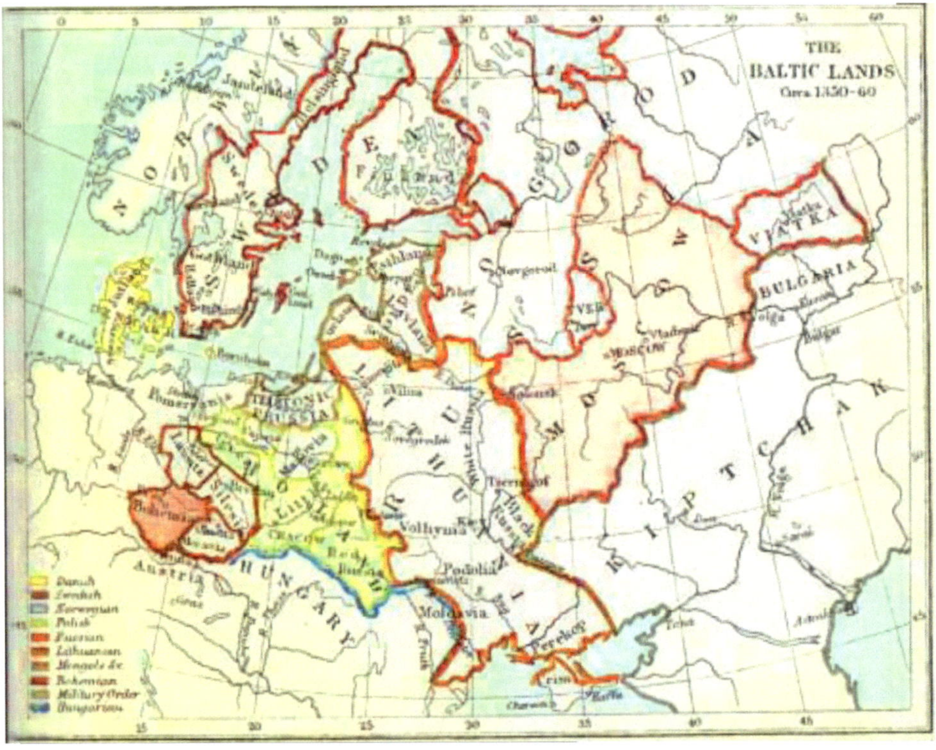
THE
BALTIC LANDS
Circa 1350-60
Danish
Swedish
Norwegian
Polish
Russian
Lithuanian
Mongols &c.
Bohemian
Military Order
Hungarian
NORWAY
SWEDEN
NOVGOROD
MOSCOW
VIATKA
BULGARIA
LITHUANIA
KIPTCHAK
HUNGARY
Austria
TEUTONIC PRUSSIA
Pomerania
Novgorod
Vilna
Tzernigof
Volhynia
Podolia
Moldavia
Perekop
Crim
Kaffa
Bulgar
Cracow
Red Russia
White Russia
Black Russia
Silesia
Bohemia
Moravia

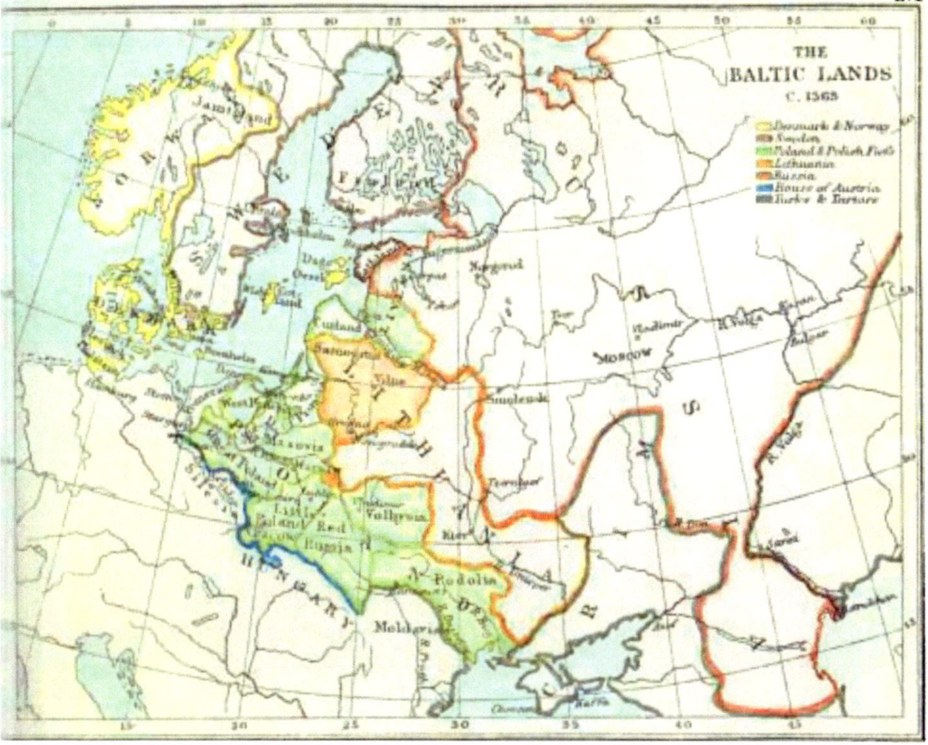
THE
BALTIC LANDS
C. 1563
Denmark & Norway
Sweden
Poland & Polish Fiefs
Lithuania
Russia
House of Austria
Turks & Tartars
Moscow
Smolensk
Vilna
Novgorod
Little Poland
Red Russia
Volhynia
Podolia
Moldavia
Silesia
Masovia
Samogitia
Curland
Dago
Osel
Gotland
Bornholm
Kaffa

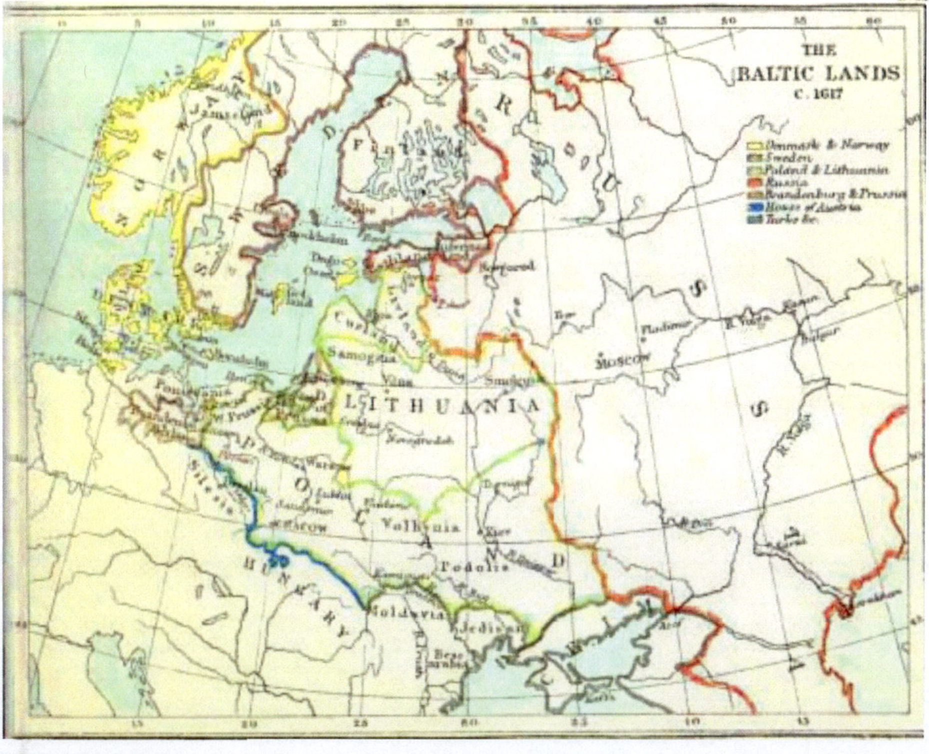
THE
BALTIC LANDS
C. 1617
Denmark & Norway
Sweden
Poland & Lithuania
Russia
Brandenburg & Prussia
House of Austria
Turks &c.
LITHUANIA
Moscow
Novgorod
Samogitia
Curland
Livland
Vilna
Pomerania
Silesia
Volhynia
Podolia
Moldavia
Jedisan
Bessarabia
Hungary
Cracow
Warsaw
Finland
Stockholm
Jamtland
Bornholm

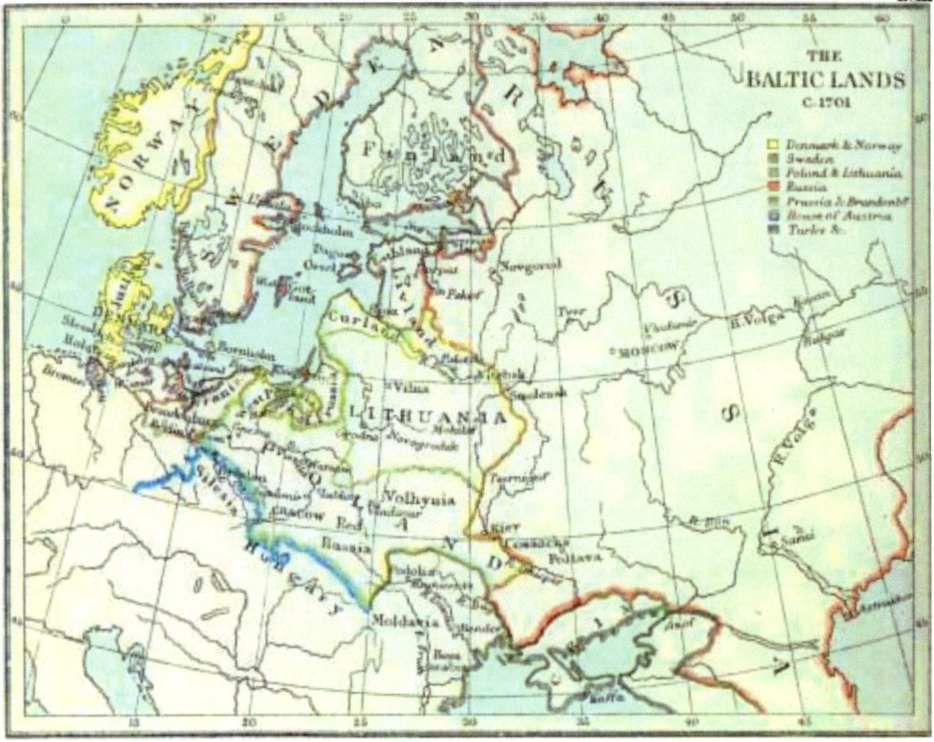
THE
BALTIC LANDS
C. 1701
Denmark & Norway
Sweden
Poland & Lithuania
Russia
Prussia & Brandenbg
House of Austria
Turks &c.
NORWAY
SWEDEN
Finland
RUSSIA
DENMARK
LITHUANIA
POLAND
Stockholm
Novgorod
MOSCOW
R. Volga
Vilna
Smolensk
Vollhynia
CRACOW
Hungary
Moldavia
Kiev
Poltava
Bender
Silesia
Curland

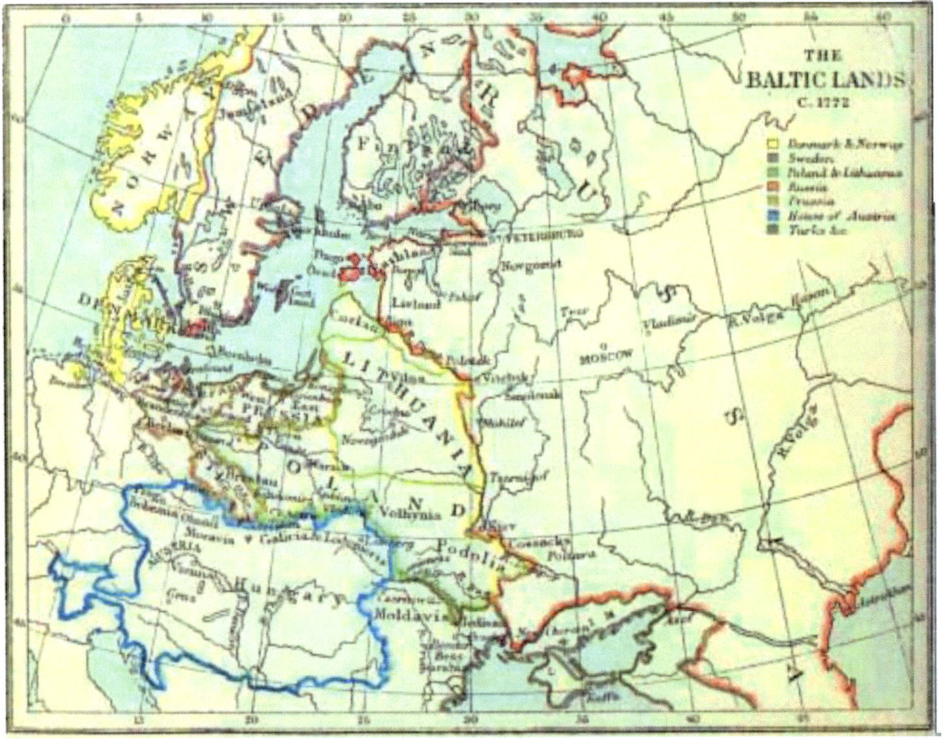
THE
BALTIC LANDS
C. 1772
Denmark & Norway
Sweden
Poland & Lithuania
Russia
Prussia
House of Austria
Turks &c.
NORWAY
SWEDEN
Finland
RUSSIA
DENMARK
LITHUANIA
POLAND
PRUSSIA
ST. PETERSBURG
Novgorod
MOSCOW
R. Volga
Vilna
Smolensk
Livland
Curland
Vollhynia
Podolia
Hungary
Moldavia
Kiev
Poltava
Moravia
Galicia
Bohemia

THE BALTIC LANDS
C. 1795

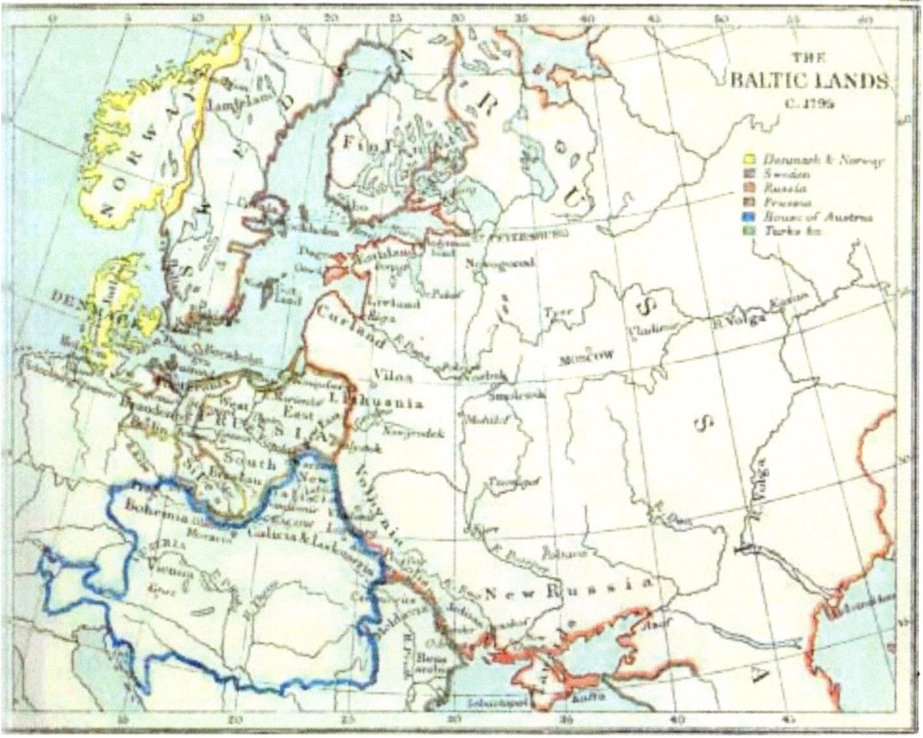

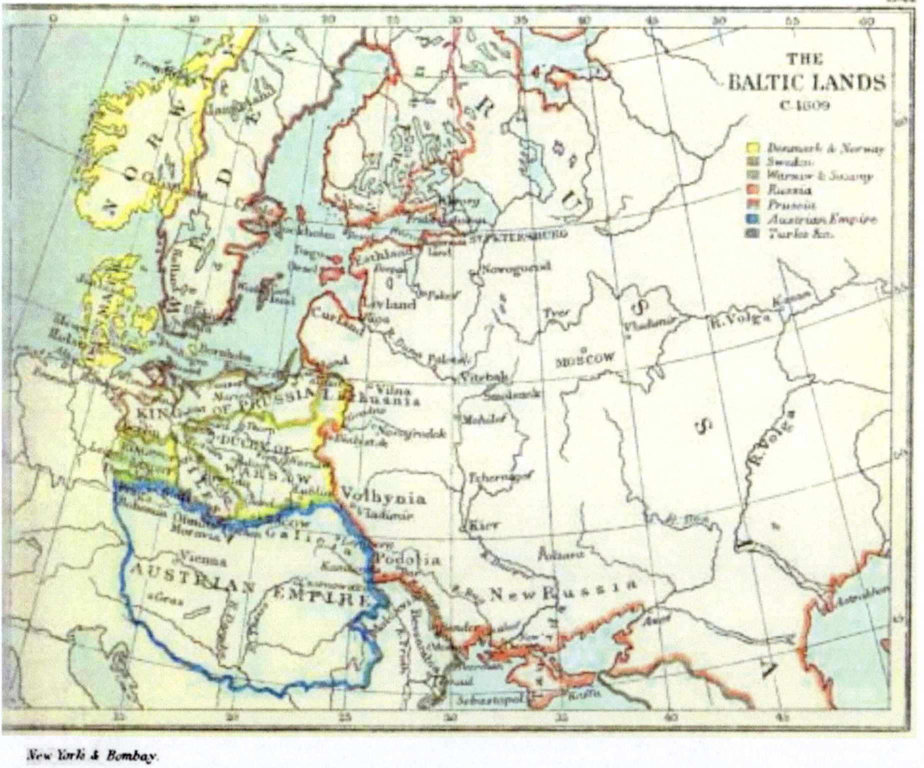

THE BALTIC LANDS
C. 1809

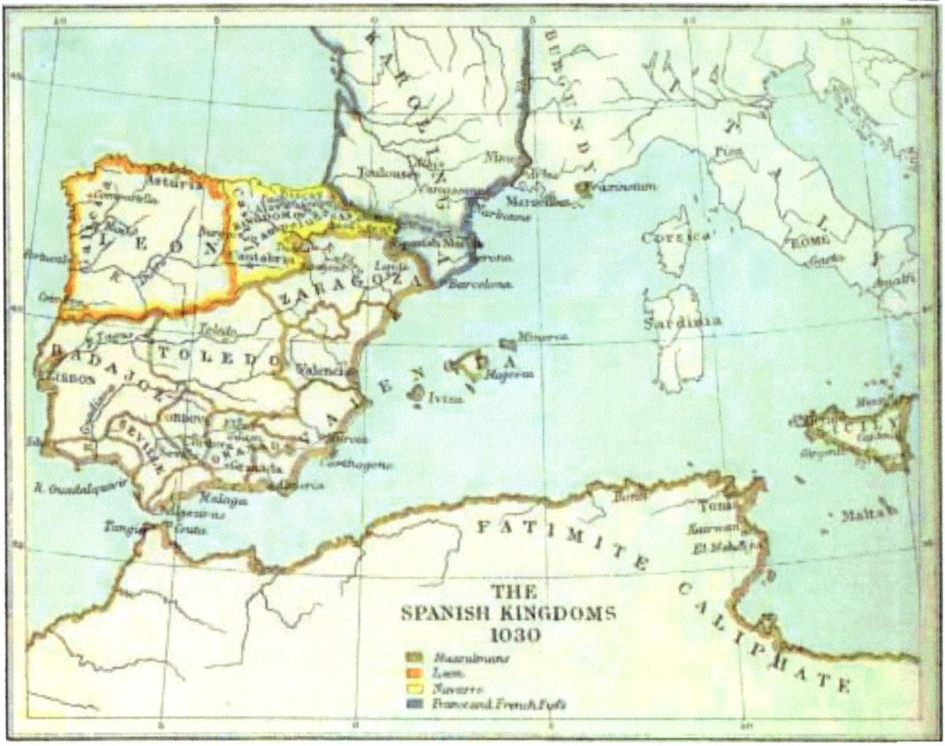
THE
SPANISH KINGDOMS
1030
Leon
Navarre
FATIMITE CALIPHATE
LEON
ZARAGOZA
TOLEDO
BADAJOZ
LISBON
SEVILLA
Asturia
Cantabria
Toledo
Valencia
Murcia
Carthagena
Granada
Malaga
Almeria
Barcelona
Marseilles
Narbonne
Toulouse
Nimes
Corsica
Sardinia
Minorca
Majorca
Iviza
ROME
Pisa
Amalfi
SICILY
Malta
Tunis
Kairwan
El Mahdiya
Bona
Tangier
Ceuta
R. Guadalquivir

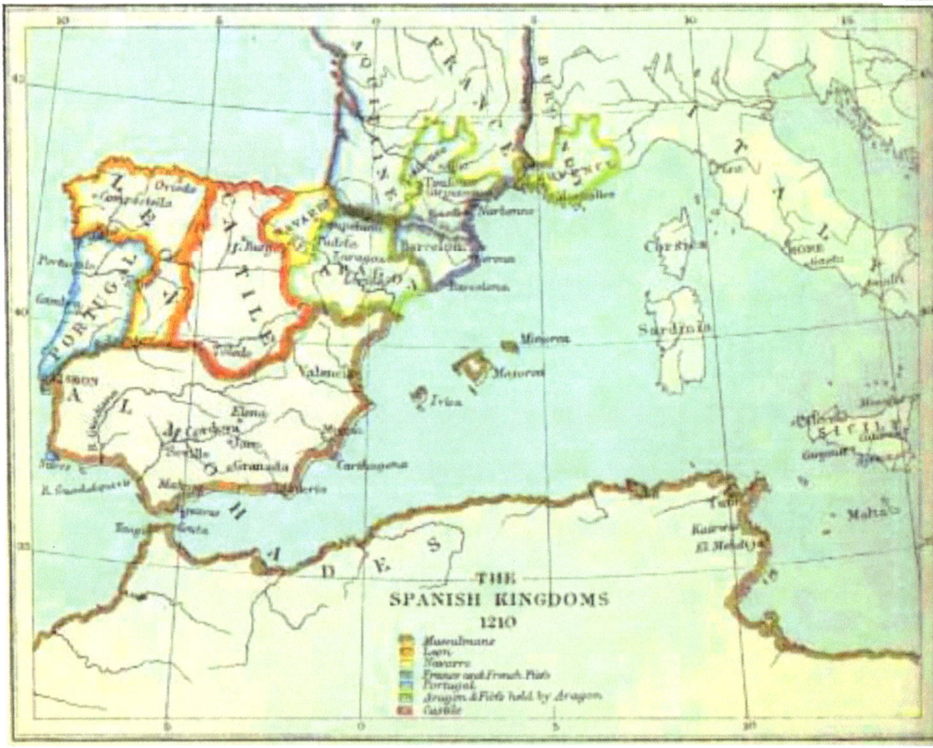
THE
SPANISH KINGDOMS
1210
Leon
Portugal
Castile
LEON
PORTUGAL
CASTILE
NAVARRE
ARAGON
FRANCE
AQUITAINE
Oviedo
Compostella
Tudela
Zaragoza
Barcelona
Narbonne
Marseilles
Valencia
Elena
Cordova
Jaen
Seville
Granada
Carthagena
Almeria
Corsica
Sardinia
Minorca
Majorca
Iviza
ROME
SICILY
Malta
Kairwan
El Mahdiya
Tangier
Ceuta
R. Guadalquivir

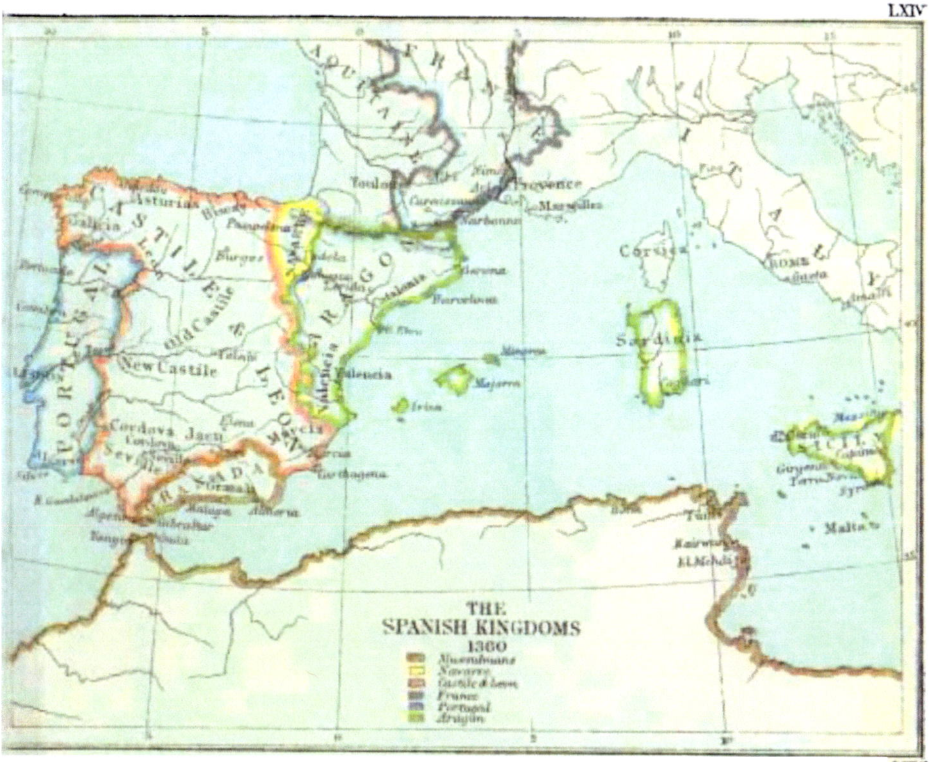
THE
SPANISH KINGDOMS
1360
Mussulmans
Navarre
Castile & Leon
France
Portugal
Aragon
CASTILE & LEON
PORTUGAL
ARAGON
GRANADA
ITALY
Asturias
Galicia
Old Castile
New Castile
Burgos
Pampelona
Toledo
Cordova
Jaen
Seville
Murcia
Valencia
Barcelona
Gerona
Carthagena
Granada
Malaga
Almeria
Gibraltar
Toulouse
Narbonne
Provence
Marseilles
Corsica
Sardinia
Cagliari
Majorca
Minorca
Iviza
ROME
SICILY
Messina
Malta
Tunis
El Mehdia

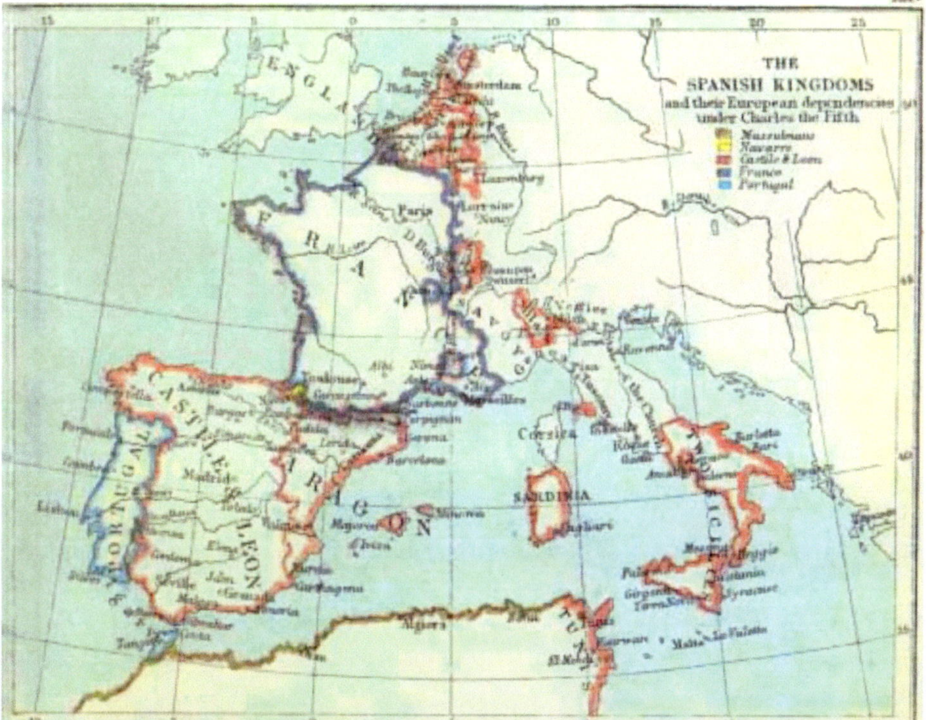
THE
SPANISH KINGDOMS
and their European dependencies
under Charles the Fifth
Mussulmans
Navarre
Castile & Leon
France
Portugal
ENGLAND
FRANCE
CASTILE & LEON
PORTUGAL
ARAGON
NAPLES
Amsterdam
Luxemburg
Paris
Nancy
Toulouse
Marseilles
Narbonne
Perpignan
Gerona
Lerida
Barcelona
Madrid
Toledo
Lisbon
Cordova
Seville
Jaen
Granada
Malaga
Almeria
Murcia
Carthagena
Majorca
Minorca
Iviza
Corsica
SARDINIA
Cagliari
Venice
Ravenna
Rome
Barletta
Bari
Amalfi
Messina
Reggio
Catania
Syracuse
Palermo
Malta
La Valetta
Algiers
Tunis
El Mehdia
Tangier
Ceuta